Superconductivity - Physics and Devices

Edited by Kim Ho Yeap and Veerendra Dakulagi

Published in London, United Kingdom

Superconductivity - Physics and Devices
http://dx.doi.org/10.5772/intechopen.1003534
Edited by Kim Ho Yeap and Veerendra Dakulagi

Contributors
Ajay Singh, Aridaman Singh Chauhan, Belqees Hassan, Bhupendra Kumar, Gedefaw Mebratie, Habtamu Anagaw, Kim Ho Yeap, Kun Liu, Morteza Heidari, Roland Harvey, Si Yuan Liang, Veerendra Dakulagi, Wei Zhou, Zhihua Qu, Zhihua Zhang

First published in London, United Kingdom, 2025 by IntechOpen
IntechOpen is the global imprint of INTECHOPEN LIMITED, registered in England and Wales, registration number: 11086078, 167-169 Great Portland Street, London, W1W 5PF, United Kingdom

For EU product safety concerns: IN TECH d.o.o., Prolaz Marije Krucifikse Kozulić 3, 51000 Rijeka, Croatia, info@intechopen.com or visit our website at intechopen.com.

British Library Cataloguing-in-Publication Data
A catalogue record for this book is available from the British Library

Superconductivity - Physics and Devices
Edited by Kim Ho Yeap and Veerendra Dakulagi
p. cm.

This title is part of the Materials Science Book Series, Volume 2
Topic: Metals and Nonmetals
Series Editor: Chonghe Li
Topic Editor: Joan Josep Roa Rovira

Print ISBN 978-1-83634-119-2
Online ISBN 978-1-83634-118-5
eBook (PDF) ISBN 978-1-83634-120-8
ISSN 3049-8856

If disposing of this product, please recycle the paper responsibly.

IntechOpen Book Series
Materials Science
Volume 2

Aims and Scope of the Series

Materials science has always occupied an extremely high position in the human development process. As we explore the oceans of stars, various industries have put forward more stringent requirements for the performance of materials, forcing us to pay more and more attention to the development of new materials. At the same time, the formation of a data-driven scientific paradigm is dramatically shortening the development cycle of new materials. The huge data generated by synergistically combining theories, high-throughput experiments, high-throughput computation, and artificial intelligence is greatly contributing to our ability to utilize materials science to solve real-world problems. The three topics of this book series - Metals and Nonmetals; Composite Materials; and Surface Science - will address important areas of advancement in materials science. There will be a range of interesting works published under these topics.

Meet the Series Editor

Prof. Chonghe Li received his Ph.D. from the Chinese Academy of Sciences in 1995. From 1995 to 2000, he worked as a researcher at the Shanghai Institute of Metallurgy, Chinese Academy of Sciences, where he also served as director of the research laboratory. In 2000, he was appointed Professor at the Institute of High-Performance Computing in Singapore, where he worked on computation and simulation in materials science until 2004. Since then, he has been a professor at the School of Materials Science and Engineering, Shanghai University, China, as well as the director of the Shanghai Specialty Casting Engineering and Technology Research Center. Prof. Li's research focuses on titanium alloys, titanium-aluminum single crystals, intermetallic compounds, theoretical calculations, alloy design, and special refractory materials. His broad scientific expertise is well recognized by the scientific community around the world. He is a member of the editorial board of the journal Metals. As an author, he has published more than 200 peer-reviewed papers, 2 books, and over 40 patents.

Meet the Volume Editors

Kim Ho Yeap holds a Bachelor of Engineering (Hons) in Electrical and Electronic Engineering from Universiti Teknologi Petronas (2004), a Master of Science in Microelectronics from Universiti Kebangsaan Malaysia (2005), and a Ph.D. in Engineering from Universiti Tunku Abdul Rahman (2011). Yeap is currently an Associate Professor at Universiti Tunku Abdul Rahman. He is a senior member of IEEE and holds Chartered Engineer registration with the UK Engineering Council, Professional Engineer registration with the Board of Engineers Malaysia, and recognition as an ASEAN Chartered Professional Engineer. An active contributor to both teaching and research, Yeap has instructed over 10 courses and conducted extensive research in electromagnetics. Since 2006, he has achieved 2 granted patents, a filed patent, and published 99 refereed journal papers, 50 international conference proceedings, 12 books, and 27 book chapters. Notably, his research contributed to designing the receiver optics for ALMA, the world's most powerful millimeter/submillimeter wave radio telescope. Throughout his career, Yeap has contributed significantly in administrative capacities, serving as the Head of the Programme for the Master of Engineering Science, Head of the Department of Electronic Engineering, and chairing the Self-Assessment Committee and the Asian Physics Olympiad competition.

Dr. Veerendra Dakulagi holds a Ph.D. in Array Signal Processing from Visvesvaraya Technological University in India, along with an M.Tech. in Power Electronics and a B.E. in Electronics and Communication Engineering. With over 15 years of experience, he has contributed to more than 60 peer-reviewed publications, including journal articles, conference papers, and book chapters. His groundbreaking work on direction-of-arrival (DOA) estimation and smart antenna systems has earned him multiple international patents. Dr. Veerendra Dakulagi is currently serving as the Professor and Head of the Department of Computer Science and Engineering (Data Science) at Guru Nanak Dev Engineering College (GNDEC) in Bidar, India, Dr. Dakulagi is also a Visiting Professor at HELP University in Malaysia (2024–2026). A recipient of numerous accolades, such as the INSA Visiting Scientist Award (2023), IEEE Bangalore Chapter Outstanding Volunteer Award (2021), and the VGST Best Researcher Award (2018), Dr. Dakulagi is recognized globally for his expertise. He has chaired international symposia and actively contributes to IEEE, advancing both research and education. His current research focuses on smart antennas and disaster-response technologies, reflecting his commitment to addressing global challenges through innovation and collaboration.

Contents

Preface **XIII**

Section 1
Foundations of Superconductivity **1**

Chapter 1 **3**
Introductory Chapter: Fundamentals of Superconductivity
by Kim Ho Yeap and Veerendra Dakulagi

Section 2
Superconducting Materials and Properties **11**

Chapter 2 **13**
Thermoelectric Properties of Superconductor Quantum Dots Hybrid Devices
by Aridaman Singh Chauhan, Bhupendra Kumar and Ajay Singh

Chapter 3 **31**
Non-Binary Cryogenic Memory Cell Equilibrium Transition Control
by Roland Harvey and Zhihua Qu

Section 3
Applications of Superconductivity **47**

Chapter 4 **49**
Applications of High-Temperature Superconductors in Microwave Devices
by Morteza Heidari

Chapter 5 **69**
Superconducting Devices: From Quantum Computing to Energy Transmission
by Belqees Hassan

Chapter 6 **81**
Superconductor Magnets
by Habtamu Anagaw and Gedefaw Mebratie

Section 4
Emerging Technologies **105**

Chapter 7 **107**
Latest Development of Superconducting Maglev
by Wei Zhou, Kun Liu, Zhihua Zhang and Si Yuan Liang

Preface

Since their discovery by Heike Kamerlingh Onnes in the early 20th century, superconductors have significantly transformed human life. Building upon their exotic and remarkable properties, such as zero electrical resistance and the ability to expel external magnetic fields, superconductors have enabled technologies that make life easier and more comfortable. These technologies span diverse fields, including transportation, biomedicine, and security.

For instance, the strong and stable magnetic fields generated by superconductors have made magnetic resonance imaging (MRI) possible. The advent of MRI has revolutionized diagnostic methods across various medical domains, including neurological, musculoskeletal, cardiovascular, abdominal, breast, pediatric, and pelvic imaging, as well as oncology. MRI enables healthcare professionals to diagnose patient health issues in detail and with high accuracy.

Another remarkable application of superconductors is in magnetic levitation technology. High-speed rail systems built on this technology allow trains to travel at speeds exceeding 400 km/h. Additionally, the negligible resistance exhibited by superconductors dramatically reduces energy losses in power transmission, making it a highly efficient option for modern energy systems.

With their myriad advantages to society, research on superconductors continues fervently.

This book provides an extensive elucidation of superconductors from multiple perspectives. To cater to a wide audience, ranging from novices to experts, it begins with the fundamental theories and concepts of superconductivity and gradually delves into advanced principles and applications. The book is divided into four main sections:

Section 1: Foundations of Superconductivity

This section begins with an introduction to the principles of superconductivity, providing a broad overview (Chapter 1).

Section 2: Superconducting Materials and Properties

This section focuses on the materials used to construct superconductors and their unique properties. Chapter 2 discusses thermoelectric and quantum effects in superconducting hybrid systems, while Chapter 3 examines advanced material applications in cryogenics and memory technology.

Section 3: Applications of Superconductivity

In this section, the practical uses of superconductors are explored. Chapter 4 highlights the application of high-temperature superconductors in microwave technologies. Chapter 5 delves into the role of superconductors in quantum computing and energy transmission, and Chapter 6 focuses on the use of superconducting magnets in technology and research.

Section 4: Emerging Technologies

The final section examines cutting-edge advancements in superconducting technology. Chapter 7 focuses on magnetic levitation (maglev) technology, showcasing its transformative potential in transportation.

This comprehensive structure ensures that readers thoroughly understand superconductors, their properties, and their far-reaching applications.

Kim Ho Yeap
Universiti Tunku Abdul Rahman,
Petaling Jaya, Malaysia

Veerendra Dakulagi
Guru Nanak Dev Engineering College,
Bidar, India

Section 1

Foundations of Superconductivity

Chapter 1

Introductory Chapter: Fundamentals of Superconductivity

Kim Ho Yeap and Veerendra Dakulagi

1. Introduction

Superconductivity refers to the phenomenon where a material transitions from a normal conducting state with finite resistance to a state of zero electrical resistance. In this superconducting state, the material also expels magnetic flux, demonstrating a hallmark of perfect diamagnetism, a property known as the Meissner effect. To induce this transition, the material's temperature must be lowered below its critical temperature, T_c.

2. Historical background

In 1908, physicist Heike Kamerlingh Onnes of Leiden University in the Netherlands achieved a groundbreaking scientific milestone by successfully liquefying helium, enabling an extremely low temperature of 4.2 K to be reached. This breakthrough opened an avenue for Onnes to investigate how materials behave at very low temperatures. In 1911, Onnes conducted experimental investigations on mercury (Hg) as it was gradually cooled to near absolute zero. To his utmost astonishment, he observed that the electrical resistance disappeared completely all of a sudden as the temperature of mercury dropped to 4.2 K. This discovery marked a monumental shift in the field of materials science since no one had ever encountered the state of zero resistivity at that time. Although Onnes referred to this phenomenon as the "supraconductivity," it is more commonly known as "superconductivity" today. Further relentless explorations by Onnes revealed that the superconductors also exhibit the Meissner effect. In 1913, Onnes was awarded the Nobel Prize in Physics for his pioneering work in low temperature physics and superconductivity.

In the subsequent decades following the discovery of superconductors, physicists attempted to explain their underlying mechanisms. However, all efforts were made to no avail. This includes the theory proposed by the London brothers at Columbia University in New York. In 1935, Fritz London and Heinz London proposed a macroscopic theory to describe the behavior of superconductors. According to the theory, surface currents are induced in a superconductor in response to an external magnetic field (this is associated with Ampère's circuital law). This means that the external magnetic field enters a superconductor at a certain penetration depth (λ_L) before it is effectively expelled. These currents generate their own magnetic fields, which oppose the applied magnetic field due to Lenz's law. The induced magnetic moment and the applied magnetic field cancel each other out, resulting in a net magnetic field of zero (i.e., the expulsion of the magnetic field) within the superconductor. Although the London theory is able to capture the essential features of superconducting behavior effectively, it fails to provide a microscopic explanation for it [1].

By the 1950s, physicist and Nobel Laureate John Bardeen of the University of Illinois at Urbana-Champaign hypothesized that the absence of resistance in a superconductor could be caused by electrons pairing up as they move. However, he was perplexed as to how this could occur since electrons are known to repel each other due to their like negative charges. In 1956, Bardeen's PhD student Leon Neil Cooper discovered that these paired electrons could be mediated by phonons, which are the vibrations of the crystal lattice. One can imagine that when an electron moves through the positively charged ions in the lattice of a superconductor, the attractive force between the negatively charged electron and the positive ions distorts the lattice. This distortion creates an accumulation of positive charge in the vicinity of the electron. Before the ions return to their original positions, a second electron is attracted to this cloud of positive charge. The attraction from this localized region of positive charge can overcome the natural repulsion between the two negatively charged electrons [2]. Today, these pairs of electrons are referred to as Cooper pairs [3]. It is worthwhile to note that the distance over which Cooper pairs couple, known as the coherence length (ξ), is relatively long. This distance is on the order of 10^{-4} cm, which is about two orders of magnitude greater than the electron mean free path (approximately 10^{-6} cm). When the ξ exceeds λ_L, London theory breaks down [1].

In 1957, John Robert Schrieffer, who was working as a postdoctoral researcher in Bardeen's group, had the epiphany that Cooper pairs could collectively form a new quantum state, which could be described using a single wave function that encompasses all electron pairs. Schrieffer's insight was the fundamental basis for the formulation of the Bardeen-Cooper-Schrieffer (BCS) theory. This theory explains that once Cooper pairs are formed, the electrons condense into a lower energy state, allowing them to move without being scattered. It also explained the Meissner effect and predicted the existence of an energy gap, which was eventually experimentally validated. The BCS theory transformed the field of condensed matter physics and earned Bardeen, Cooper, and Schrieffer the Nobel Prize in Physics in 1972.

3. The BCS theory

In the BCS theory, Cooper pairs are formed when two electrons bind via phonon interactions as they move through the lattice [4]. The wave function for a Cooper pair is given as

$$\Psi_{\text{pair}}(r_1, r_2) = \frac{1}{\sqrt{2}}[\psi(r_1)\psi(r_2) - \psi(r_2)\psi(r_1)]. \tag{1}$$

This antisymmetric wave function arises due to the Pauli exclusion principle, as the individual electrons (fermions) in the Cooper pair must obey Fermi-Dirac statistics. Since Cooper pairs are bosonic, multiple pairs are allowed to occupy the same quantum state without violating Pauli exclusion principle. This collective behavior enables the coherent flow of electrons without scattering.

The BCS ground state describes the superconducting state in which electrons near the Fermi surface form Cooper pairs. It can be expressed as [5].

$$|\Psi_{\mathrm{BCS}}\rangle = \prod_{\mathbf{k}} \left(u_{\mathbf{k}} + v_{\mathbf{k}} c^{\dagger}_{\mathbf{k}\uparrow} c^{\dagger}_{-\mathbf{k}\downarrow} \right) \bigg| 0\rangle \tag{2}$$

where $u_{\mathbf{k}}$ and $v_{\mathbf{k}}$ are probability amplitudes that depend on the momentum $\mathbf{k}$ (indicating whether a state is occupied or unoccupied by a Cooper pair), and they are related to the energy gap (Δ) and Fermi level. The operators $c^{\dagger}_{\mathbf{k}\uparrow}$ and $c^{\dagger}_{-\mathbf{k}\downarrow}$ are creation operators for electrons with momentum $\mathbf{k}$ and spin $\uparrow$ and momentum $-\mathbf{k}$ and spin $\downarrow$, respectively. The product is taken over all electron momenta $\mathbf{k}$ so that the pairs occupy opposite momenta ($\mathbf{k}$, $-\mathbf{k}$). The state $|0\rangle$ denotes the vacuum state where electrons are absent.

The energy of the system is given by

$$E(\mathbf{k}) = \sqrt{(\epsilon_{\mathbf{k}} - \mu)^2 + \Delta^2} \tag{3}$$

where $\epsilon_{\mathbf{k}}$ is the single-particle energy for momentum $\mathbf{k}$ and μ is the chemical potential. Electrons form Cooper pairs by condensing into a discrete energy level, resulting in the creation of an energy gap (Δ) at the Fermi surface. The energy gap Δ represents the binding energy per electron in a Cooper pair. Since there are two electrons in a Cooper pair, the minimum energy required to break them apart is 2Δ. The energy gap Δ is temperature T dependent and can be computed from [6–8].

$$\ln(\tilde{\Delta}) = -2\int_0^{\infty} \left(E^2 + \tilde{\Delta}^2\right)^{-1/2} \left\{1 + \exp\left[\left(\pi/\gamma_E \tilde{T}\right)\left(E^2 + \tilde{\Delta}^2\right)^{1/2}\right]\right\}^{-1} dE \tag{4}$$

where $\tilde{\Delta} = \Delta(T)/\Delta(0)$, $\tilde{T} = T/T_c$, and γ_E = 1.781 is the Euler's constant.

The critical temperature T_c can be approximated by the following relation:

$$T_c \approx \frac{1.14 h\omega_D}{2\pi k_B} \exp\left(-\frac{1}{N(0)V}\right) \tag{5}$$

where h = 6.626 × 10^{-34} Js is Planck's constant, k_B = 1.380649 × 10^{-23} J/K is the Boltzmann's constant, ω_D is the Debye frequency related to the phonon spectrum in the material, $N(0)$ is the density of states at the Fermi level, and V denotes the attractive interaction potential between electrons mediated by phonons.

The BCS theory also explains the Meissner effect in a microscopic manner. The formation of Cooper pairs and their macroscopic quantum state, which gives rise to nonresistant supercurrents, prevent the external magnetic field from penetrating the superconductor beyond a characteristic distance, known as the London penetration depth λ_L. To be more exact, λ_L is similar to the skin depth δ in a normal conductor, where it measures the distance at which the field attenuates to e^{-1} of its initial amplitude. Hence, the magnetic field does not cease abruptly at λ_L; rather, it continues to tail off beyond this depth. This means that the external magnetic field penetrates the superconductor up to λ_L, where it is exponentially damped by induced supercurrents. These supercurrents generate a magnetic field that continuously cancels the incident magnetic field within the bulk of the superconductor, ensuring that the interior remains free of magnetic flux. The London penetration depth λ_L can be estimated using the following expression:

$$\lambda_L = \sqrt{\frac{m}{\mu_0 n_s e^2}} \tag{6}$$

where m is the electron mass, μ_0 is the permeability of free space, n_s is the superconducting electron density, and e is the electron charge.

Upon inspecting the London equation in (7) below, it can be seen that the supercurrent $\mathbf{j}$ is directly proportional to the magnetic vector potential $\mathbf{A}$

$$\mathbf{j} = -\frac{n_s e^2}{m}\mathbf{A} \tag{7}$$

The magnetic vector potential $\mathbf{A}$ can be derived based on the Biot-Savart law as

$$\mathbf{A}(\mathbf{r}) = \frac{\mu_0}{4\pi}\int \frac{\mathbf{j}(\mathbf{r}')}{|r - \mathbf{r}'|} dV' \tag{8}$$

where $\mathbf{r}$ and $\mathbf{r}'$ are, respectively, the position vectors of the observation and source points. The curl of $\mathbf{A}$ gives the magnetic flux density $\mathbf{B}$

$$\mathbf{B} = \nabla \times \mathbf{A} \tag{9}$$

4. The complex conductivity

When a material is in a superconducting state (i.e., its temperature is below T_c), thermal energy and incident radiation can disrupt Cooper pairs. When a pair is broken by these external factors, the resulting unpaired electrons are referred to as excited quasiparticles. These quasiparticles behave like normal electrons with well-defined momenta. Since the binding energy of a Cooper pair is 2Δ, the absorption of incident radiation with a frequency exceeding $\frac{2\Delta}{h}$ is required to break the pairs. This frequency threshold is known as the gap frequency.

According to the BCS theory, the energy levels of the electrons just above and below the Fermi energy E_F are separated by the energy gap 2Δ when the material enters the superconducting state [9–12]. Intuitively, no quasiparticle states are expected within this energy gap. Contrarily, however, experimental measurements have shown that this is not the case at all. The quasiparticle states discovered within the energy gap are referred to as intragap states in [13–17].

To account for the presence of intragap states, Noguchi et al. [16, 17] extended the BCS theory to develop a modified complex conductivity model for superconductors. Unlike the real gap energy predicted by the traditional BCS theory, however, Noguchi et al. introduce a complex gap energy, $\Delta = \Delta_1 + j\Delta_2$, where both Δ_1 and Δ_2 are real values, with Δ_1 aligning with the real gap energy defined in Eq. (4). Noguchi et al. selected niobium (Nb) as the superconducting material, with a critical temperature T_c of 9.2 K, an energy gap at 0 K of $2\Delta(0) = 3.05$ meV, and a normal-state conductivity $\sigma_n = 1.57 \times 10^7$ S/m [10]. In actual Nb films, the imaginary component of the energy gap Δ_2 is observed to be 10^{-4} of the real component Δ_1 [16, 17]. The surface resistance of superconductors estimated using this new formulation closely matches experimental measurements. By extending the formulation in [18], the complex conductivity can be expressed as [10]

$$\begin{aligned}\frac{\sigma_1 - j\sigma_2}{\sigma_n} &= \frac{2}{\hbar\omega}\int_{\Delta_1}^{\infty} \frac{[f(E_r) - f(E_r+\hbar\omega)]\left(E_r^{\,2}+\Delta^2+\hbar\omega E_r\right)}{\left(E_r^{\,2}-\Delta^2\right)^{1/2}\left[(E_r+\hbar\omega)^2-\Delta^2\right]^{1/2}}dE_r \\ &+ \frac{1}{\hbar\omega}\int_{-\Delta_1}^{-\Delta_1+\hbar\omega} \frac{[1-2f(E_r-\hbar\omega)]\left(E_r^{\,2}+\Delta_1^{\,2}-\hbar\omega E_r\right)}{\left(E_r^{\,2}-\Delta_1^{\,2}\right)^{1/2}\left[(E_r-\hbar\omega)^2-\Delta_1^{\,2}\right]^{1/2}}dE_r \\ &- \frac{j}{\hbar\omega}\int_{-\Delta_2}^{0} \frac{[1-2f(\Delta_1+jE_i)]\left[(\Delta_1+jE_i)\left(\Delta_1^{\,2}+jE_i-\hbar\omega\right)+\Delta^2\right]}{\left[(\Delta_1+jE_i)-\Delta^2\right]^{1/2}\left[(\Delta_1+jE_i-\hbar\omega)^2-\Delta^2\right]^{1/2}}dE_i \\ &+ \frac{j}{\hbar\omega}\int_{-\Delta_2}^{0} \frac{[1-2f(\Delta_1+jE_i)]\left[(\Delta_1+jE_i)\left(\Delta_1^{\,2}+jE_i+\hbar\omega\right)+\Delta^2\right]}{\left[(\Delta_1+jE_i)-\Delta^2\right]^{1/2}\left[(\Delta_1+jE_i+\hbar\omega)^2-\Delta^2\right]^{1/2}}dE_i\end{aligned} \tag{10}$$

where E_r and E_i are the real and imaginary parts, respectively, of the complex quasiparticle excitation energy E; $=\frac{h}{2\pi}$ is the reduced Planck's constant; ω is the angular frequency; and σ_n is the normal state conductivity. The real part of the integrals represents quasiparticles excited thermally and by the fields above the gap frequency f_g, while the imaginary part represents the contribution from Cooper pairs. Notably, the last two integrals on the right-hand side of (10) account for the contribution of quasiparticles and Cooper pairs in the intragap states. The function f that describes the Fermi-Dirac statistics is expressed in terms of E as

$$f(E) = \frac{1}{1 + exp\,(E/k_B T)} \tag{11}$$

5. Types of superconductors

In general, superconductors can be classified into two categories: those that adhere to the principles of the BCS theory and those that do not, known as nonconventional superconductors.

Superconductors that comply with the BCS theory are low-temperature superconductors (LTSs), which can be further divided into Type I and Type II. The key distinction between Type I and Type II superconductors is that Type I is characterized by the complete expulsion of magnetic fields, exhibiting a single critical magnetic field. In contrast, Type II superconductors allow partial penetration of magnetic fields and possess two critical magnetic fields, enabling them to maintain their superconducting state under higher magnetic field strengths. Examples of Type I superconductors include lead (Pb), mercury (Hg), and tin (Sn), while examples of Type II superconductors include niobium (Nb) and iron-based superconductors.

Nonconventional superconductors are those that do not fully adhere to the BCS theory and exhibit some exotic properties. High-temperature superconductors (HTSs) are a type of nonconventional superconductors. These types of superconductors typically possess a critical temperature (T_c) above 30 K, with some exceeding the boiling point of liquid nitrogen (77 K). Examples of HTSs include yttrium barium copper oxide, bismuth strontium calcium copper oxide, and mercury barium calcium copper oxide. There are also superconductors with low T_c that do not comply with the BCS theory. Examples of these nonconventional superconductors include heavy fermion superconductors, organic superconductors, and iron-based superconductors.

6. Conclusion

Since their discovery in the early twentieth century, superconductors have been a fervent area of research, primarily due to their remarkable ability to exhibit zero resistance and expel external magnetic fields. The growing understanding of superconductivity over the years has paved the way for various practical applications that underpin the significance of superconductors in modern technology. These applications span diverse fields, including biomedicine, security, and transportation.

For instance, superconducting magnets are employed in magnetic resonance imaging (MRI) devices because of their strong and stable magnetic fields. By leveraging their zero-resistance feature, superconductors also enhance electrical grids, improving power transmission efficiency. Additionally, superconductors are used to construct magnetic levitation trains (Maglev) for high-speed rail systems. By levitating above the ground, Maglev trains can reach speeds of more than 400 km/h. Furthermore, superconductors play an integral role in the operation of particle accelerators and fusion reactors, which are vital for advancing scientific research. Superconducting circuits also hold significant potential in the field of quantum computing, presenting exciting prospects for future technology. There is no doubt that advancements in our understanding of superconductivity can enhance the standard of living for humanity.

Author details

Kim Ho Yeap[1*] and Veerendra Dakulagi[2]

1 Universiti Tunku Abdul Rahman, Kampar, Perak, Malaysia

2 Guru Nanak Dev Engineering College, Bidar, India

*Address all correspondence to: yeapkh@utar.edu.my

References

[1] Matick RE. Transmission Lines for Digital and Communication Networks. US: McGraw-Hill; 1969

[2] Yeap KH, Yeong KC, Tham CY, Nisar H. Analysis of energy loss in superconducting waveguides. In: Ponnusamy V, Zaman N, Low TJ, Amin AHM, editors. Biologically-Inspired Energy Harvesting through Wireless Sensor Technologies. US: IGI Global; 2016

[3] Cooper LN. Bound electrons pairs in a degenerate Fermi gas. IEEE Transactions on Microwave Theory and Techniques. 1956;**104**:1189-1190

[4] Bardeen J, Cooper LN, Schrieffer JR. Theory of superconductivity. Physical Review. 1957;**108**:1175-1204

[5] Duzer TV, Turner CW. Principles of Superconductive Devices and Circuits. Upper Saddle River, NJ: Prentice Hall; 1999

[6] Kautz PL. Picosecond pulses on superconducting striplines. Journal of Applied Physics. 1978;**49**: 308-314

[7] Yeap KH, Tham CY, Yeong KC, Lim EH. Full wave analysis of normal and superconducting microstrip transmission lines. Frequenz Journal of RF-Engineering and Telecommunications. 2010;**64**:56-66

[8] Yeap KH, Tham CY, Yeong KC, Woo HJ. Wave propagation in lossy and superconducting circular waveguides. Radioengineering Journal. 2010;**19**: 320-325

[9] Wengler MJ. Submillimeter waves detection with superconducting tunnel junctions. Proceedings of the IEEE. 1992; **80**:1810-1826

[10] Yeap KH, Teh JSM, Nisar H, Yeong KC, Hirasawa K. Attenuation in superconducting rectangular waveguides. Frequenz Journal of RF-Engineering and Telecommunications. 2015;**69**:111-117

[11] Yeap KH, Ong SS, Nisar H, Lai KC, Ng CA. Attenuation in superconducting circular waveguides. Advanced Electromagnetics. 2016;**5**:34-38

[12] Tucker JR, Feldman MJ. Quantum detection at millimeter wavelengths. Reviews of Modern Physics. 1985;**57**: 1055-1113

[13] Mitrovic B, Rozema LA. On the correct formula for the lifetime broadened superconducting density of states. Journal of Physics. Condensed Matter. 2008;**20**:015215

[14] Noguchi T, Suzuki T, Endo A, Tamura T. Contribution of the imaginary part of the superconducting gap energy on the SIS tunneling current. Physica C. 2009;**469**:1585-1588

[15] Noguchi T, Suzuki T, Tamura T. Subgap tunneling current at low temperature in Nb/Al-AlN/Nb SIS junctions. IEEE Transactions on Applied Superconductivity. 2011;**21**:756-759

[16] Noguchi T, Naruse M, Sekimoto Y. RF conductivity and surface impedance of a superconductor taking into account the complex superconducting gap energy. Physics Procedia. 2012;**36**: 318-323

[17] Noguchi T, Naruse M, Sekimoto Y. Contribution of quasiparticles in the

subgap states to the surface impedance of superconductors. IEEE Transactions on Applied Superconductivity. 2013;**23**: 1501404

[18] Mattis DC, Bardeen J. Theory of the anomalous skin effect in normal and superconducting metals. Physics Review. 1958;**111**:412-417

Section 2

Superconducting Materials and Properties

Chapter 2

Thermoelectric Properties of Superconductor Quantum Dots Hybrid Devices

Aridaman Singh Chauhan, Bhupendra Kumar and Ajay Singh

Abstract

The study of supercurrent transport in hybrid superconductor quantum dot mesoscopic devices has been a prominent area of research for several decades due to its promising applications in nanoelectronics. This review provides a theoretical perspective on Josephson transport within these hybrid superconductor quantum dot systems. We begin with a concise overview of essential theoretical concepts, including Bardeen-Cooper-Schrieffer (BCS) mean-field theory, Josephson effects, quantum dots, and Andreev bound states. Initially, we examine the Josephson and thermal transport through uncorrelated double quantum dots (single-level) arranged in a T-shaped side-coupled configuration and situated between two Bardeen-Cooper-Schrieffer (BCS) superconducting leads, modeled using the single-impurity Anderson model's Hamiltonian and solved via Green's equation of motion technique. Subsequently, we review the current-phase relationship and the corresponding energy-phase relation of Andreev bound states (ABSs) for different quantum dot energy levels relative to the Fermi level and interdot-hopping parameter at absolute zero temperature. Finally, we explore the thermoelectric transport properties across the junction, analyzing the behavior of the Josephson supercurrent and quasi-particle current through the quantum dots under varying interdot-hopping, thermal biasing, and quantum dot energy level positions.

Keywords: quantum dots, superconductivity, Josephson transport, π-junction, single-impurity Anderson model, Andreev bound states (ABS), BCS theory, Green's equation of motion technique, thermoelectric transport, hybrid superconductor devices

1. Introduction

A superconductor-quantum dot-superconductor (SC-QD-SC) Josephson junction can be made by comprising a quantum dot between two Bardeen-Cooper-Schrieffer (BCS) superconducting electrodes. This configuration allows a direct current (DC) Josephson supercurrent to traverse the junction even without an applied voltage, primarily because the supercurrent is highly sensitive to the phase difference between the superconductors. Nanofabrication techniques have revolutionized the ability to fabricate devices in which superconducting and/or normal leads are connected to quantum dots (QDs). These hybrid devices, known as mesoscopic devices, exhibit distinct characteristics owing to their nanoscale dimensions, which range from 100 Å to 1 μm, and electronic properties markedly different from those of bulk materials [1–4]. Quantum dots possess discrete energy levels, which can be modulated either by adjusting the gate voltage or by altering the quantum dot's size [5, 6]. In a Josephson junction where the insulation barrier is replaced with a quantum dot, both quasi-particle (single electrons) tunneling and Cooper pairs tunneling play critical roles in facilitating charge transport. By carefully controlling these parameters, one can effectively influence the behavior and characteristics of the supercurrent flowing through the junction. Utilizing quantum dots enables precise regulation of the current that flows through these junctions, which was not feasible in the case of standard Josephson junctions. Additionally, numerous researchers have investigated the charge transport characteristics of Josephson junctions incorporating double quantum dots. By manipulating the quantum dots' properties, such as their energy levels and coupling strengths, it is possible to achieve a high degree of control over the supercurrent and other electronic properties of the junction. These studies have provided valuable insights into the complex interactions and mechanisms governing charge transport in these advanced quantum systems. Recent comprehensive reviews and detailed analysis on the electronic transport properties of single along with double QD-based Josephson junctions can be found in Refs. [7–12]. These reviews offer in-depth analyses and discussions on the mechanisms of charge transport, highlighting the unique behaviors and phenomena observed in these systems.

Conversely, the thermal transport characteristics of Josephson junctions (SIS) and those incorporating quantum dots have not been extensively studied, primarily due to the limited temperature range within which they function. However, there has been a growing interest in the thermal transport characteristics of these Josephson junctions in recent years. Although research in this area is still emerging, some recent studies have begun to investigate the thermal transport behavior of QD-based Josephson junctions, particularly when both leads are superconducting in nature. These pioneering studies are starting to illuminate the mechanisms of heat transfer in these complex systems, contributing to a more comprehensive understanding of their thermal behavior.

The thermoelectric transport properties of the above-discussed QD-based Josephson junctions, driven by both phase and temperature variations, can be analyzed by examining two types of currents: quasi-particle current and Cooper pair current. A thermal gradient across the junction causes a current induced by quasi-particles outside the superconducting gap to flow across the junction. In the SC-QD-SC configuration, the quasi-particle current is the primary contributor to thermal transport. The pair current, also referred to as the Josephson supercurrent, traverses the junction without any voltage or temperature difference between the superconducting (SC) leads. This Josephson current is influenced by the phase difference between the SC leads. By understanding these distinct currents and their respective roles, one can gain valuable insights into the complex transport phenomena occurring in quantum dot-based Josephson junctions (**Figure 1**).

In this review, we present an analysis of the thermoelectric transport characteristics in a system comprising uncorrelated double quantum dots, arranged in a T-shaped configuration and coupled to two superconducting leads. In this setup, the primary quantum dot (QD_1) is directly connected to the superconducting leads, while the secondary quantum dot (QD_2) is connected to the primary dot but not to the leads [7, 8]. To investigate the thermal transport properties of this T-shaped DQD

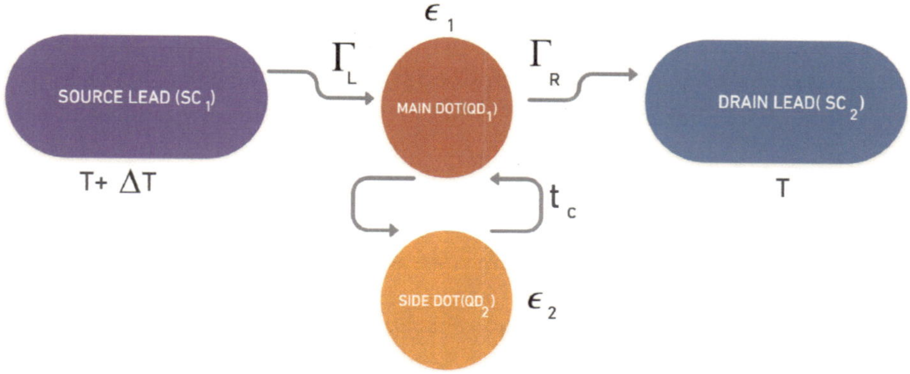

Figure 1.
Uncorrelated T-shaped/side-coupled Double Quantum Dot (DQD) Josephson junction.

Josephson junction, we utilized the Green's equation of motion technique. Initially, we examined how the interdot-hopping strength (t_c) and quantum dot's energy levels (ε_1 and ε_2) affect the supercurrent. Subsequently, we calculated the total current, which includes both the quasi-particle current along with the Josephson supercurrent (due to Cooper pairs), for various temperature differences (ΔT). By doing so, we aim to gain a comprehensive understanding of the thermal as well as electronic transport in this intricate quantum system.

2. BCS mean-field theory

BCS superconductors, named after Bardeen, Cooper, and Schrieffer, who developed the theory in 1957, are a class of materials that exhibit zero electrical resistance below a critical temperature. This phenomenon occurs due to the formation of Cooper pairs, where electrons with opposite spins and momenta pair up as a result of their interactions with lattice vibrations (phonons). This pairing leads to the creation of a superconducting energy gap (Δ) in the electronic density of states below the critical temperature (T_C) BCS superconductors are characterized by macroscopic quantum coherence, enabling phenomena such as persistent currents and the Josephson effect, which are critical for applications in quantum computing and advanced sensing technologies. These materials include elemental superconductors like lead and niobium, as well as more complex compounds such as cuprates and iron-based superconductors, which can exhibit high-temperature superconductivity. Practical applications of BCS superconductors range from medical imaging devices like MRI machines to particle accelerators and highly efficient power transmission systems, making them essential for both fundamental research and technological innovations. These junctions consist of two superconductors separated by a thin insulating barrier. The Josephson effect allows a supercurrent—a current with zero electrical resistance—to flow across the junction, influenced by the phase difference between the left and the right superconducting wavefunctions on either side of the barrier. This phenomenon has significant implications for both fundamental physics and practical applications in technology [13–15].

It was a known fact in superconducting physics that Fermi sea gets unstable due to the electronic scattering in the quantum states $|\mathbf{k}, \uparrow\rangle$ and $|-\mathbf{k}, \downarrow\rangle$. [14] Thereafter, Bardeen, Cooper, and Schrieffer formulated an ansatz for the new superconducting ground state [13, 15, 16]. The basic idea was that electrons with opposite spin and momenta would interact attractively due to distortion by the surrounding lattice (electron-phonon interaction) and form Cooper pairs, and the ground state is a superposition of multiple states that are built up of such pairs. The ansatz reads

$$|\psi_{\mathrm{BCS}}\rangle = \prod_{\mathbf{k}} \left(u_{\mathbf{k}} + v_{\mathbf{k}} c^{\dagger}_{\mathbf{k}\uparrow} c^{\dagger}_{-\mathbf{k}\downarrow} \right) |0\rangle, \tag{1}$$

where $|0\rangle$ corresponds to the vacuum state and $u_{\mathbf{k}}$, $v_{\mathbf{k}}$ are arbitrary complex coefficients.

The variational ansatz used above gives us an approximation of the many-particle superconducting ground state, but it does not allow us to predict thermodynamic properties.

Despite being a variational ansatz, it lacks much insight into the conceptual character of the approximation. Hence, instead of that superconductors at finite temperatures within mean-field theory must be considered, that provides a brand new perspective on the gap equation.

We start again from the Hamiltonian

$$H = \sum_{\mathbf{k}\sigma} \xi_k c^{\dagger}_{\mathbf{k}\sigma} c_{\mathbf{k}\sigma} + \frac{1}{N} \sum_{\mathbf{k}\mathbf{k}'} V_{\mathbf{k}\mathbf{k}'} c^{\dagger}_{\mathbf{k}\uparrow} c^{\dagger}_{-\mathbf{k},\downarrow} c_{-\mathbf{k}',\downarrow} c_{\mathbf{k}'\uparrow}. \tag{2}$$

A *mean-field approximation* replaces the products of two operators, say A and B according to

$$AB \cong \langle A \rangle B - \langle A \rangle \langle B \rangle + A \langle B \rangle. \tag{3}$$

$$H_{\mathrm{BCS}} = \sum_{\mathbf{k}\sigma} \xi_k c^{\dagger}_{\mathbf{k}\sigma} c_{\mathbf{k}\sigma} + \frac{1}{N} \sum_{\mathbf{k}\mathbf{k}'} V_{\mathbf{k}\mathbf{k}'} \left(\left\langle c^{\dagger}_{\mathbf{k}\uparrow} c^{\dagger}_{-\mathbf{k},\downarrow} \right\rangle c_{-\mathbf{k}',\downarrow} c_{\mathbf{k}'\uparrow} - \left\langle c^{\dagger}_{\mathbf{k}\uparrow} c^{\dagger}_{-\mathbf{k},\downarrow} \right\rangle \langle c_{-\mathbf{k}',\downarrow} c_{\mathbf{k}'\uparrow} \rangle + c^{\dagger}_{\mathbf{k}\uparrow} c^{\dagger}_{-\mathbf{k},\downarrow} \langle c_{-\mathbf{k}',\downarrow} c_{\mathbf{k}'\uparrow} \rangle \right). \tag{4}$$

We define

$$\Delta_{\mathbf{k}} := -\frac{1}{N} \sum_{\mathbf{k}'} V_{\mathbf{k}\mathbf{k}'} \langle c_{-\mathbf{k}',\downarrow} c_{\mathbf{k}'\uparrow} \rangle \tag{5}$$

so that

$$\Delta^{*}_{\mathbf{k}} = -\frac{1}{N} \sum_{\mathbf{k}'} V_{\mathbf{k}\mathbf{k}'} \left\langle c^{\dagger}_{\mathbf{k}'\uparrow} c^{\dagger}_{-\mathbf{k}',\downarrow} \right\rangle. \tag{6}$$

$$H_{\mathrm{BCS}} = \sum_{\mathbf{k}\sigma} \xi_{\mathbf{k}} c^{\dagger}_{\mathbf{k}\sigma} c_{\mathbf{k}\sigma} - \sum_{\mathbf{k}} \Delta^{*}_{\mathbf{k}} c_{-\mathbf{k},\downarrow} c_{\mathbf{k}\uparrow} - \sum_{\mathbf{k}} \Delta_{\mathbf{k}} c^{\dagger}_{\mathbf{k}\uparrow} c^{\dagger}_{-\mathbf{k},\downarrow} + \text{const.} \tag{7}$$

The constant has little bearing on the derivation that follows. Given that H_{BCS} is obviously bilinear with respect to $c, c^{\dagger}$, it characterizes an effective non-interacting system. Unexpectedly, terms like cc and $c^{\dagger}c^{\dagger}$ can be found in H_{BCS} that do not conserve the electronic number. As a result, we anticipate that H_{BCS}'s eigenstates lack conserved electron numbers. This is peculiar because it is never observed that states with different quantities of electrons can superimpose (**Figure 2**).

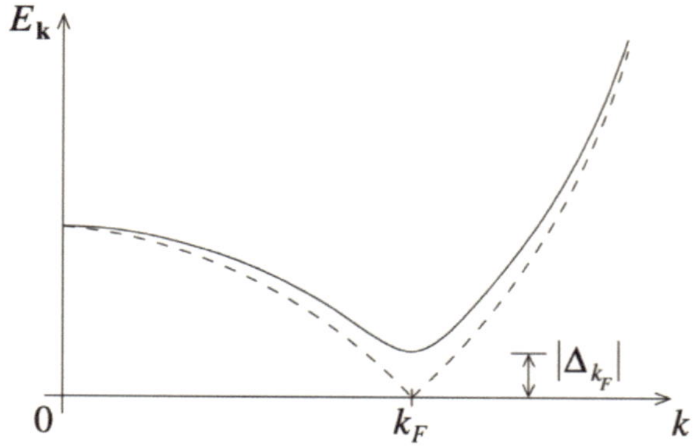

Figure 2.
Energy dispersion relation in superconductors [11].

3. Transport in superconductor-quantum dot systems

3.1 Josephson effects

A supercurrent flows between two superconductors separated by an extremely thin insulating barrier, creating a Josephson junction, as part of the quantum mechanical phenomenon known as the Josephson effect (**Figure 3**) [18].

3.1.1 DC Josephson effect

The DC Josephson effect occurs in the absence of an external voltage. In this scenario, a direct current (DC) can flow across the junction due to the quantum mechanical tunneling of Cooper pairs, which are pairs of electrons with opposite spins and momenta. The supercurrent (I_{SC}) depends on the phase difference (ϕ) between the superconducting wavefunctions on either side of the junction, following the relation $I_{SC} = I_C sin\phi$ where I_C is the critical current—the maximum supercurrent that can flow through the junction without resistance.

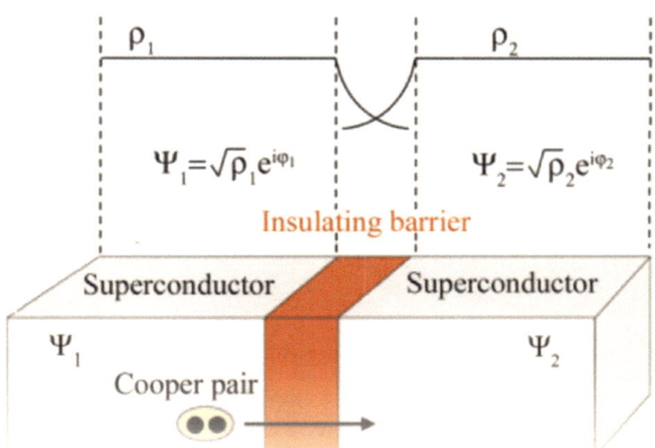

Figure 3.
Cooper pair tunneling in a standard Josephson junction. Figure from ref. [17].

3.1.2 AC Josephson effect

In contrast, the AC Josephson effect occurs when a constant voltage is applied across the Josephson junction, resulting in an alternating current (AC) supercurrent. The frequency of this AC current is directly proportional to the applied voltage, described by the relation $f = \frac{2eV}{h}$, where e is the elementary charge and h is Planck's constant. The applied voltage induces a time-dependent phase difference, leading to an oscillating supercurrent. This effect is notable for its ability to produce high-frequency electromagnetic radiation and is crucial in various technological applications (**Figure 4**).

3.2 Quantum dots

Quantum dots are semiconductor nanostructures renowned for their unique quantum mechanical properties, particularly the size-dependent energy levels they exhibit due to electron confinement in all three dimensions. These nanoscale structures effectively act as artificial atoms, with energy levels that can be finely tuned by altering their size [19–22]. The first hint of zero-dimensional quantum confinement was given by Ekimov in 1980 [23]. Larger quantum dots possess closely spaced energy levels, while smaller ones feature larger energy gaps between these levels. This size-dependent quantum confinement not only influences their electronic structure but also dictates their optical properties. Quantum dots can absorb light at specific wavelengths determined by their size and emit light at shorter wavelengths, a phenomenon crucial for applications in high-resolution displays, advanced lighting systems, and precise biomedical imaging techniques. Understanding and controlling these quantum properties of energy levels and size are pivotal for harnessing quantum dots in a wide array of technological innovations, from quantum computing to next-generation solar cells and beyond.

3.3 Andreev bound states

The Andreev bound states are fundamental phenomena in superconducting systems, particularly at the interface between a quantum dot or another mesoscopic

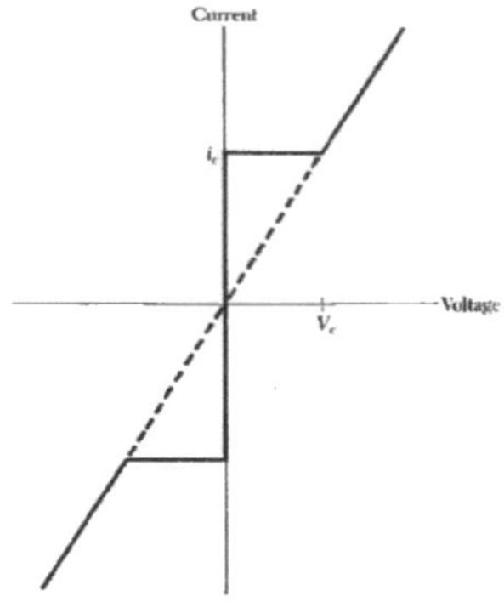

Figure 4.
*I-V Characteristic plot of a Josephson junction, DC current flow under negligible bias up to a critical supercurrent I_c—this comes under the DC Josephson effect. As the voltage bias exceeds V_c the junction has a finite resistance, but the current develops a periodic component of frequency $\omega = 2eV/\hbar$—this is termed as AC Josephson effect (**Figure 5**).*

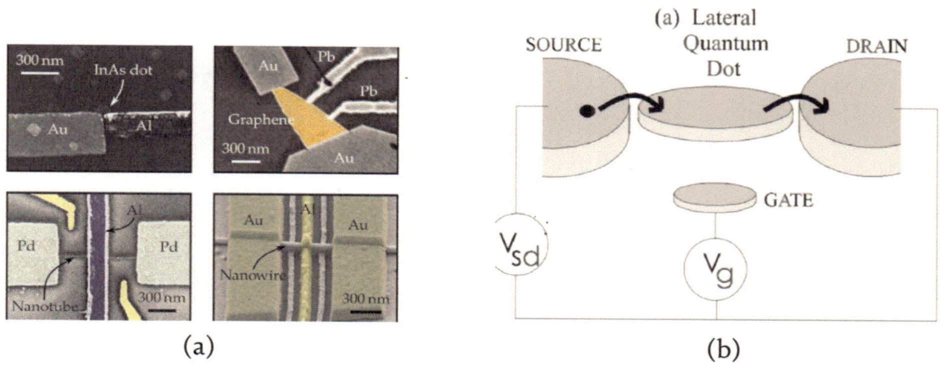

Figure 5.
(a) Various quantum dots coupled to a superconductor: Recent fabrication advancements made it possible to sandwich QD with normal or/and superconducting leads and study the electronic transport across such junctions (Fig. adopted from ref. [2]). (b) Schematic of a QD, in the shape of a disk, connected to source and drain terminals by tunneling junctions and to a gate terminal by a capacitor-like junction (Fig. adopted from ref. [17]).

structure and superconducting leads. These states originate from multiple Andreev reflection (MAR) processes, which occur when an electron with energy below the superconducting energy gap (Δ) interacts with the superconductor. During MAR, the incident electron pairs with a second electron of opposite momentum within the superconductor, forming a Cooper pair, while simultaneously creating a hole in the quantum dot with nearly identical momentum but opposite velocity. The energy conservation laws dictate that the energies of the incoming electron and the resulting hole relative to the Fermi level of the superconductor are symmetric around the Fermi level. This symmetry allows for the formation of discrete bound states within the superconducting energy gap, known as Andreev bound states. These bound states manifest as two-level systems: one state resides below the Fermi level (sub-gap state), and the other resides above it (above-gap state) (**Figure 6**) [9, 10, 24].

4. Formulation

4.1 Hamiltonian

Using the second quantized BCS Hamiltonian discussed above and an extended Anderson impurity model, we write the Hamiltonian corresponding to the electronic transport across the junction, after which the thermoelectric transport properties of the T-shaped DQD Josephson junction are analyzed [7, 11, 25, 26]:

$$\hat{H} = \hat{H}_{\text{leads}} + \hat{H}_{\text{QD}} + \hat{H}_{\text{tunnel}} + \hat{H}_{\text{interdot-hopping}} \tag{8}$$

where

$$\hat{H}_{\text{leads}} = \sum_{k\sigma,\alpha} \epsilon_{k\alpha} c^{\dagger}_{k\sigma,\alpha} c_{k\sigma,\alpha} - \left(\sum_{k\alpha} \Delta_{\alpha} c^{\dagger}_{k\uparrow,\alpha} c^{\dagger}_{-k\downarrow,\alpha} + h.c. \right) \tag{9}$$

$$\hat{H}_{\text{QD}} = \sum_{i=1}^{2} \sum_{\sigma} \epsilon_i d^{\dagger}_{i\sigma} d_{i\sigma} \tag{10}$$

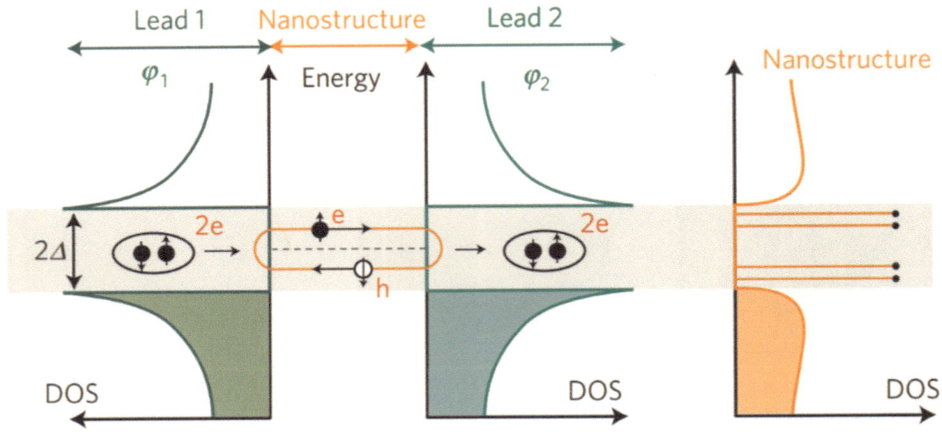

Figure 6.
Andreev bound states in Josephson junctions [12].

$$\hat{H}_{\text{tunnel}} = \sum_{k\sigma,\,\alpha} V_{k,\alpha} c^{\dagger}_{k\sigma,\alpha} d_{1\sigma} + h.c. \tag{11}$$

$$\hat{H}_{\text{interdot-hopping}} = \sum_{\sigma} t_c \left(d^{\dagger}_{1\sigma} d_{2\sigma} + h.c. \right) \tag{12}$$

where h.c. denotes the Hermitian conjugate.

The term $\hat{H}_{\text{leads}}$ describes the left and right superconducting (SC) terminals ($\alpha \in L, R$). The first term represents free electrons in the SC leads, where $c^{\dagger}_{k\sigma,\alpha}$ ($c_{k\sigma,\alpha}$) are the creation (annihilation) operators for electrons with spin σ, wave vector $\vec{k}$, and energy $\epsilon_{k,\alpha}$. The second term in $\hat{H}_{\text{leads}}$ corresponds to the BCS part, providing information about Cooper pair interactions and the temperature-dependent SC gap.

$$\Delta_\alpha(T_\alpha) = \Delta_0 \tanh\left\{ 1.74 \sqrt{\left(\frac{k_B T_c}{k_B T_\alpha} - 1 \right)} \right\} \tag{13}$$

given that $\Delta_0 = |\Delta_0| e^{i\phi_\alpha}$ is the superconducting (SC) gap and ϕ_α is the phase of the SC leads, T_c represents the critical temperature of the SC leads, T_α is the temperature of the SC leads, and k_B is the universal Boltzmann constant.

The Hamiltonian term $\hat{H}_{\text{QD}}$ describes the side quantum dot (QD_2) and the main quantum dot (QD_1). Here, $d^{\dagger}_{i\sigma}$ ($d_{i\sigma}$) denotes the fermionic creation (annihilation) operator for electrons with spin σ, and $n_{d_i\sigma} = d^{\dagger}_{i\sigma} d_{i\sigma}$ is the corresponding number operator. The energies of QD_1 and QD_2 are represented by ϵ_i ($i = 1, 2$). For simplicity, onsite intradot-Coulomb interactions U_i are assumed to be negligible.

The tunneling part of the Hamiltonian, $\hat{H}_{\text{tunnel}}$, describes the interaction between the energy level of the main QD and the SC leads with an interaction strength $V_{k,\alpha}$. Additionally, symmetric coupling is assumed, with the coupling strength of QD_1 to the left and right leads being $V_{k1,L} = V_{k1,R}$.

The final term, $\hat{H}_{\text{interdot-hopping}}$, describes the electronic interaction between the two quantum dots with a hopping strength t_c. Note that the SC leads and QD_2 do not directly interact.

The Bogoliubov-Valatin transformation is employed in order to diagonalize the BCS term.

$$c_{k\uparrow} = u_k^* \beta_{k\uparrow} + v_k \beta_{-k\downarrow}^\dagger \tag{14}$$

$$c_{-k\downarrow}^\dagger = u_k \beta_{-k\downarrow}^\dagger - v_k^* \beta_{k\uparrow} \tag{15}$$

$$\begin{aligned}
\hat{H} = &\sum_{k\alpha} E_{k\alpha}\left(\beta_{k\uparrow\alpha}^\dagger \beta_{k\uparrow\alpha} + \beta_{-k\downarrow\alpha}^\dagger \beta_{-k\downarrow\alpha}\right) \\
&+\sum_{k\alpha}\left(V_{k\alpha} u_k^* \beta_{k\uparrow\alpha}^\dagger d_{1\uparrow} + V_{k\alpha} u_k^* \beta_{-k\downarrow\alpha}^\dagger d_{1\downarrow}\right) \\
&+\sum_{k\alpha}\left(V_{k\alpha}^* u_k d_{1\uparrow}^\dagger \beta_{k\uparrow\alpha} + V_{k\alpha}^* u_k d_{1\downarrow}^\dagger \beta_{-k\downarrow\alpha}\right) \\
&+\sum_{k\alpha} V_{k\alpha} v_k \left(\beta_{-k\downarrow\alpha} d_{1\uparrow} - \beta_{k\uparrow\alpha} d_{1\downarrow}\right) \\
&+\sum_{k\alpha} V_{k\alpha}^* v_k^* \left(d_{1\uparrow}^\dagger \beta_{-k\downarrow\alpha} - d_{1\downarrow}^\dagger \beta_{k\uparrow\alpha}\right) \\
&+\epsilon_1\left(d_{1\uparrow}^\dagger d_{1\uparrow} + d_{1\downarrow}^\dagger d_{1\downarrow}\right) \\
&+\epsilon_2\left(d_{2\uparrow}^\dagger d_{2\uparrow} + d_{2\downarrow}^\dagger d_{2\downarrow}\right) \\
&+t_c\left(d_{1\uparrow}^\dagger d_{2\uparrow} + d_{1\downarrow}^\dagger d_{2\downarrow} + d_{2\uparrow}^\dagger d_{1\uparrow} + d_{2\downarrow}^\dagger d_{1\downarrow}\right)
\end{aligned} \tag{16}$$

where $E_{k\alpha}$ is nothing but $\sqrt{\epsilon_{k,\alpha}^2 + |\Delta_\alpha|^2}$ which is the excitation energy of the quasi-particles in the superconducting terminals. The coefficients v_k and u_k can then be rewritten as:

$$|u_k|^2 = \frac{1}{2}\left(1 + \frac{\epsilon_{k,\alpha}}{\sqrt{\epsilon_{k,\alpha}^2 + |\Delta_\alpha|^2}}\right) \tag{17}$$

$$|v_k|^2 = \frac{1}{2}\left(1 - \frac{\epsilon_{k,\alpha}}{\sqrt{\epsilon_{k,\alpha}^2 + |\Delta_\alpha|^2}}\right) \tag{18}$$

Green's EOM method (EOM) is used to solve the effective Hamiltonian discussed above [7, 25, 26]. The T-shaped double quantum dot Josephson junction system's thermoelectric transport properties can be computed using the Retarded Green's function (single particle) of the main quantum dot QD_1, which is defined using Zubarev's notation [25]:

$$\langle\langle d_\sigma(t); d_\sigma^\dagger(0)\rangle\rangle = -i\theta(t)\left\langle \left\{d_\sigma(t), d_\sigma^\dagger(0)\right\}_+\right\rangle \tag{19}$$

The above retarded single particle Green's function in k-space should satisfy EOM written below:

$$\omega\langle\langle d_\sigma | d_\sigma^\dagger\rangle\rangle_\omega = \left\langle \left\{d_\sigma, d_\sigma^\dagger\right\}_+\right\rangle + \langle\langle \left[d_\sigma, \hat{H}\right]_- | d_\sigma^\dagger\rangle\rangle_\omega \tag{20}$$

Using the above equation and Hamiltonian, we get a closed system of six coupled equations, which are solved to obtain the final expression of Retarded Green's function (single particle) of the main QD with spin $\sigma = \uparrow$ (calculations are skipped):

$$G^r_{11}(\omega) = \left\langle\left\langle d_{1\uparrow} | d^\dagger_{1\uparrow} \right\rangle\right\rangle = \frac{\omega + \epsilon_1 - \frac{t_c^2}{\omega+\epsilon_2} - I_1}{\left(\omega + \epsilon_1 - \frac{t_c^2}{\omega+\epsilon_2} - I_1\right)\left(\omega - \epsilon_1 - \frac{t_c^2}{\omega-\epsilon_2} - I_2\right) - (I_3)^2} \tag{21}$$

The diagonal in the aforementioned Green's function is represented by the values I_1 and I_2, whereas the self-energy's off-diagonal portion, or I_3, relates to the induced pairing caused by the direct coupling of the main QD and SC leads.

The expressions for I_1, I_2, and I_3 are

$$I_1 = I_2 = -\sum_{\alpha \in \{L,R\}} \left(\frac{\Gamma_\alpha \omega}{\sqrt{\Delta_\alpha^2 - \omega^2}} \theta(\Delta - |\omega|) + i \left(\frac{\Gamma_\alpha \omega}{\sqrt{\omega^2 - \Delta_\alpha^2}} \theta(|\omega| - \Delta) \right) \right) \tag{22}$$

$$I_3 = -\sum_{\alpha \in \{L,R\}} \left(\frac{\Gamma_\alpha \Delta_\alpha}{\sqrt{\Delta_\alpha^2 - \omega^2}} \theta(\Delta - |\omega|) + i \left(\frac{\Gamma_\alpha \Delta_\alpha}{\sqrt{\omega^2 - \Delta_\alpha^2}} \theta(|\omega| - \Delta) \right) \right) \tag{23}$$

4.2 Supercurrent and quasi-particle current

In Ref. [7], a comprehensive derivation detailing the phase-driven Josephson supercurrent and thermal-driven quasi-particle (QP) current is provided. The expressions for these currents, as directly utilized in this study and also referenced in [7, 8, 26], are explicitly stated as follows:

$$I_{SC} = \frac{e}{h} \sum_\sigma \int d\omega \frac{\Gamma_\alpha^2 \Delta_\alpha^2}{\omega^2 - \Delta_\alpha^2} f(\omega) Im \left[\frac{-1}{A(\omega)} \right] \tag{24}$$

$$I_{QP} = \frac{e}{h} \sum_\sigma \int d\omega \frac{df(\omega)}{dT} \Gamma_\alpha Re\,(\rho(\omega)) Im \left[-G^r_{d\sigma}(\omega) \right] \Delta T \tag{25}$$

Here, $f(\omega)$ is the Fermi-distribution function, $\rho(\omega)$ is the generalized density of states, ΔT is the thermal bias (in linear regime), and $A(\omega)$ is the denominator of Retarded Green's function of the main QD.

In a standard SC-QD-SC junction, it is known that the supercurrent consists of two parts: the continuous part that is because of the electrons that lie above and below the SC gap region, while the discrete part lies within the SC gap. The expression above for the supercurrent, when divided into two integrals as discussed above, can then be solved by direct integration in the case of the continuous part and by solving the poles of the Green's function in the case of the discrete part. The major contribution comes from the discrete part of the supercurrent in the case of SC-QD-SC junctions. The poles of the Green function that lie within the SC gap are nothing but Andreev bound states, which are responsible for the electronic transport. In the case of a single QD sandwiched between two superconducting leads, two poles representing a pair of ABS are obtained: one below the Fermi energy and one above (equal but opposite signs).

5. Thermoelectric transport in T-shaped DQD Josephson junction

In this section, we reproduce and review the numerical results and provide analysis for the T-shaped DQD Josephson junction, focusing on the transport properties of uncorrelated quantum dots [7, 16]. Throughout our investigation, we use the superconducting energy gap (Δ_0) at absolute zero temperature as our energy reference, where Δ_0 is expressed in milli-electron volts (meV).

In the case of the T-shaped DQD junction, we get two pairs of Andreev bound states, unlike the case discussed above when solving for the poles of the single particle Retarded Green function because of the presence of a side quantum dot. Moreover, an SC-QD-SC junction behaves as a π junction when the energy level of the main QD is at the Fermi level($E = 0$), and the supercurrent reverses its direction when $\theta = \pi$ condition is satisfied as the ABS' become degenerate and their energies cross the Fermi level, leading to resonant tunneling that causes the supercurrent to enhance and reach a maxima when this condition holds. When QD_2 is considered, the supercurrent no longer behaves as a π junction for $\epsilon_1 = 0$ and I_{sc} VS θ plot shows a sinuous line shape instead of a discrete jump at $\theta = \pi$. Thus, a question arises about what relation or condition the parameters must satisfy in a T-shaped DQD junction for it to behave as a π-junction.

When the relation $\epsilon_1\epsilon_2 = t_c^2$ is satisfied in a T-shaped junction, one pair of Andreev bound States (ABS) crosses at precisely $E = 0$ and becomes degenerate, causing the junction to behave as a π junction and resulting in the reversal of the supercurrent direction at $\theta = \pi$.

In **Figure 7(a)**, the current exhibits a sudden jump at $\theta = \pi$ when the condition $\epsilon_1\epsilon_2 = t_c^2$ is met. Analytically, this condition can be derived from the expression $A(\omega) = 0$, where $A(\omega)$ represents the denominator of the retarded single particle Green's function of the main quantum dot QD_1. When the ABS intersect at $E = 0$, solving $A(\omega = 0)$ and subsequently setting $\theta = \pi$ straightforwardly yields $\epsilon_1\epsilon_2 = t_c^2$. This signifies that one pair of bound states becomes precisely degenerate at the Fermi energy, leading to a noticeable jump in the supercurrent at $\theta = \pi$. Furthermore, the condition $\epsilon_1\epsilon_2 = t_c^2$ also corresponds to achieving the maximum critical current in the system.

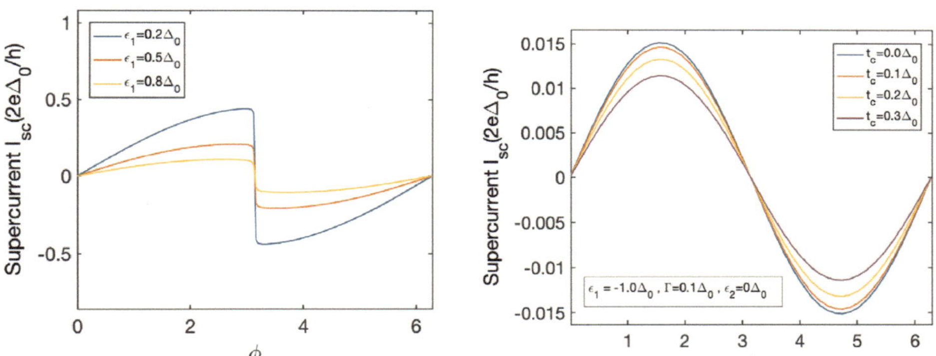

(a) (b)

Figure 7.
(a) I_{sc} vs. ϕ for various ϵ_1 under the condition $\epsilon_1\epsilon_2 = t_c^2$ and $t_c^2 = \Gamma = 0.1$ and (b) I_{sc} vs. (ϕ) for different values of t_c at absolute zero temperature; $\Gamma = 0.1\Delta_0$, $\epsilon_1 = -1.0\Delta_0$, and $\varepsilon_2 = 0\Delta_0$.

As the interdot-hopping parameter increases, **Figure 7(b)** illustrates that the supercurrent is notably suppressed. This suppression arises because increased coupling between QD_1 and QD_2 facilitates electron tunneling into QD_2. This process leads to destructive interference between the two transport channels, ultimately diminishing the magnitude of the supercurrent. Consequently, when $t_c > 0$, the system ceases to function as an ideal transmitting channel, as it does in the case of $t_c = 0$ (SC-QD-SC junction). A deeper insight into the reduction of supercurrent can be gained by examining how the energy levels of the quantum dots (QDs) split due to the interdot-hopping parameter. Specifically, when $t_c \neq 0$, the energy levels of the QDs split into two distinct levels, denoted as $\overline{\epsilon}_i = \epsilon_i \pm t_c$. As the interdot-hopping strength increases, these split energy levels $\overline{\epsilon}_i$ move further away from the origin (Fermi level). Hence, on increasing t_c, ABS drifts further away from the Fermi surface (refer **Figure 8(b)**). This shift results in a reduction of the supercurrent, as the effective coupling between the QDs and the superconducting leads diminishes with increasing interdot-hopping, thereby influencing the overall transport characteristics of the system.

Figure 8 describes the behavior of the two pairs of Andreev bound states as discussed above for different tunable parameters. **Figure 8(a)** shows the ABS's

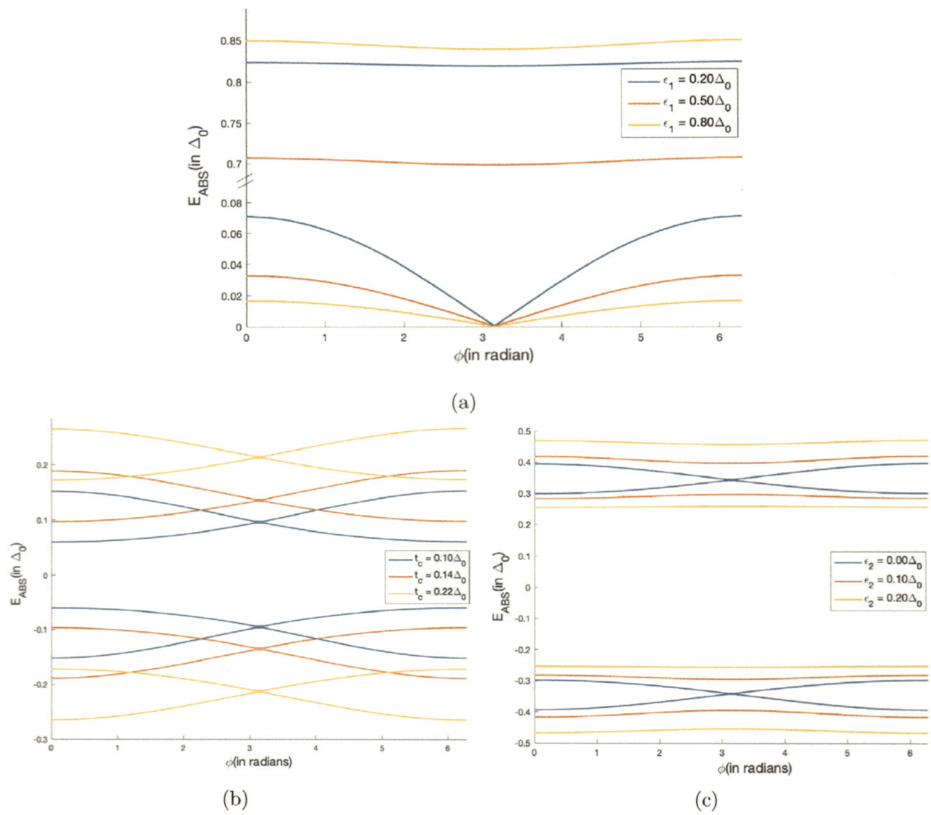

Figure 8.
(a)E_{ABS} VS ϕ is plotted for various ϵ_1 values with ϵ_1 and ϵ_2 being related by $\epsilon_1\epsilon_2 = t_c^2$ with $t_c^2 = \Gamma = 0.1$(Only the portion above the Fermi level is shown lying in the interval $[0, \Delta]$) (b)E_{ABS} VS ϕ is plotted for various t_c with $\epsilon_1 = \epsilon_2 = 0$ and $\Gamma = 0.1\Delta_0$ (c)E_{ABS} VS ϕ for different ϵ_2 with $\epsilon_1 = 0$, $t_c = 0.36\Delta_0$ and $\Gamma = 0.1\Delta_0$.

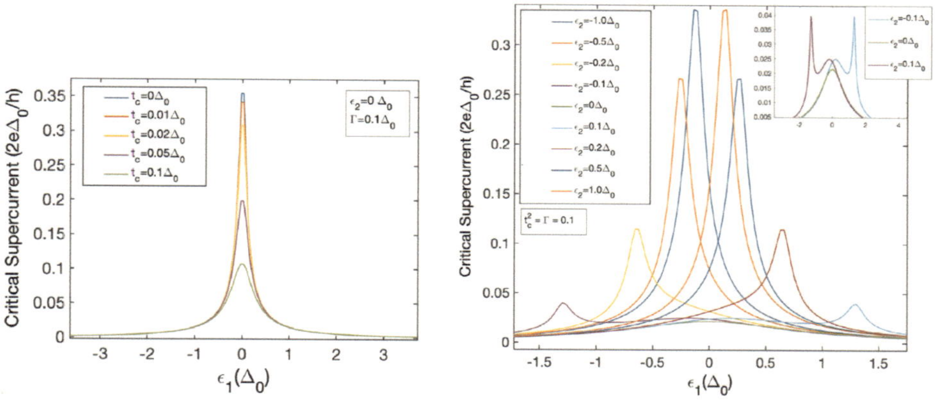

(a) (b)

Figure 9.
(a) Critical supercurrent versus ϵ_1 is plotted for various t_c. (b) Critical supercurrent versus ϵ_1 is plotted for various ϵ_2. For given t_c, I_c shows a local maxima at $\epsilon_1 = 0$ with the side quantum dot's energy at the Fermi level($\epsilon_2 = 0$) (inset). Graphs reproduced from [11].

corresponding to the parameters of **Figure 7(a)**. Clearly, one can observe that a single pair of ABS becomes degenerate at the Fermi level (as the ABS are symmetric about $E = 0$, only the positive region of the superconducting gap is shown) and at $\theta = \pi$ when $\epsilon_1\epsilon_2 = t_c^2$ holds. In **Figure 8(b)**, one can observe that the two pairs of ABS get further away from the Fermi surface with increasing magnitude of interdot-hopping strength (refer to the explanation of **Figure 7(b)**). In **Figure 8(c)**, ABS cross each other only at $\epsilon_2 = 0$ and $\theta = \pi$ but they remain non-degenerate and drift apart for increasing values of ϵ_2. This leads to the suppression of the magnitude of the supercurrent when the energy of the side quantum dot is at the Fermi level. (Refer inset of **Figure 9(b)**).

Figures 9 and **10** represent the critical value (maximum value over the 2π period of the superconductor's phase) of the supercurrent varied with different junction parameters and are discussed below.

In **Figure 9(a)**, we observe a symmetric peak between the critical supercurrent and ϵ_1 for various values of the interdot-hopping parameter while fixing $\epsilon_2 = 0$. The figure being symmetric about the Fermi energy ($E = 0$) represents the electron-hole symmetry in the superconducting leads. Moreover, the peak occurs at $\epsilon_1 = 0$ as the electrons flow through the system smoothly when the energy level ϵ_1 aligns with the Fermi energy. However, when the side quantum dot QD_2 is taken into account with some finite value of interdot-hopping parameter and its energy level $\epsilon_2 = 0$, electrons tunneling through QD_1 tend to also tunnel into QD_2. This scenario leads to destructive interference between the two paths, resulting in a reduced supercurrent despite ϵ_2 being close to the Fermi level. QD_2 effectively acts as an impurity, scattering incident electrons or Cooper pairs.

In **Figure 9(b)**, the critical supercurrent is plotted against ϵ_1 for various values of ϵ_2. Constructive interference begins to manifest as ϵ_2 moves away from the Fermi energy. An additional peak emerges, and the original peak at $\epsilon_1 = 0$ becomes asymmetric. The reason for this remains the same: when the condition $\epsilon_1\epsilon_2 = t_c^2$ holds, a pair of Andreev bound states aligns with the Fermi level, facilitating electronic transport.

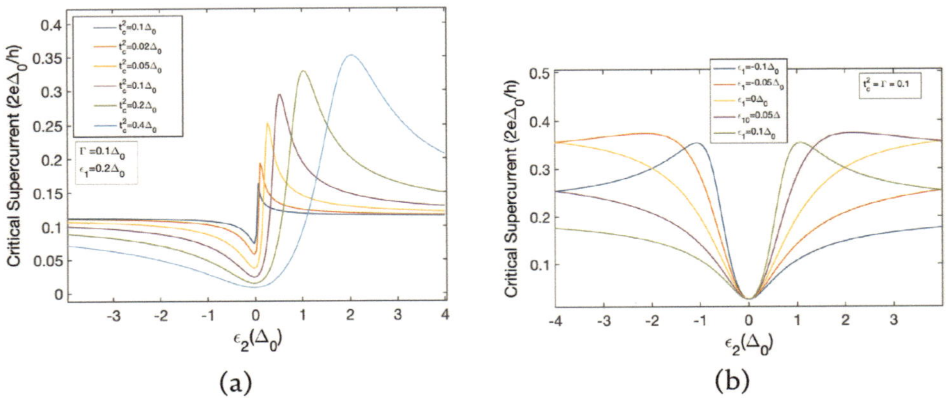

Figure 10.
(a) Critical supercurrent versus ϵ_2 is plotted for various t_c. A line shape that resembles a typical Fano line (asymmetric in nature) when ϵ_1 is non-zero with a local minima at $\epsilon_2 = 0$ and a peak that is formed according to $\epsilon_1\epsilon_2 = t_c^2$. (b) Critical supercurrent versus ϵ_2 is plotted for various ϵ_1. A minima at $\epsilon_2 = 0$ and a maxima at $\epsilon_2 = \frac{t_c^2}{\epsilon_1}$. Graphs reproduced from [11].

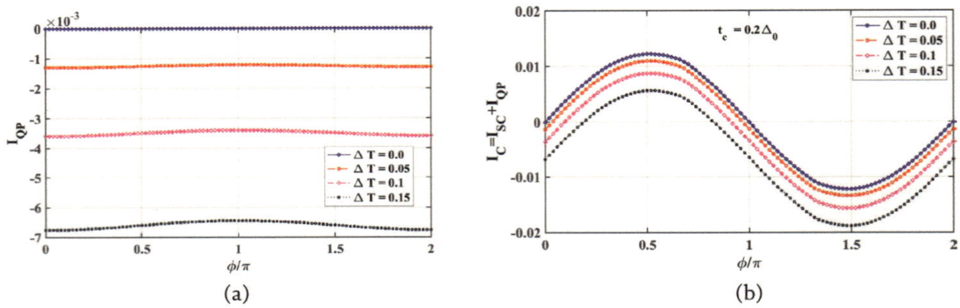

Figure 11.
(a) I_{qp} VS ϕ is plotted for various values of ΔT (b) $I_{qp} + I_{sc}$ VS ϕ is plotted for various values of ΔT in the linear regime; $\Gamma = 0.1\Delta_0$, $t_c = 0.2\Delta_0$, $\epsilon_{d_1} = -1.0\Delta_0$, and $\epsilon_{d_2} = 0$.

In **Figure 10(a)**, the critical supercurrent versus ϵ_2 is plotted for various t_c. We observe a Fano-like line shape that can be adjusted by varying t_c. The Fano line shape, often observed in spectroscopy and other areas of physics, describes an asymmetric resonance profile resulting from the system's response due to the constructive and destructive interference between a discrete resonant state and a continuum of states. This interaction leads to a distinctive resonance that is not symmetric like the Lorentzian profile typically seen in simple resonance phenomena. In the T-shaped junction, the superconducting DOS represents the continuum, while the hybrid energy level of QD_1 and QD_2 represents the discrete quantum state. This interference modifies the line shape from a symmetric Lorentzian peak to an asymmetric profile, reflecting the complex interaction dynamics. The interference effects, both constructive and destructive, are more pronounced with higher t_c values as seen in **Figure 10 (a)**. Specifically, the Fano dip minima remain fixed at $\epsilon_2 = 0$ for smaller critical supercurrents, while the local maxima shift further away from the Fermi level as the critical current magnitude increases. In **Figure 10(b)**, we depict the critical value of I_{sc} versus ϵ_2 for different values of ϵ_1. The Fano maxima can be adjusted, causing the peak's position to shift from the right side when $\epsilon_2 > 0$ to the left side when $\epsilon_2 < 0$, depending on the variation of ϵ_1. This plot again clearly illustrates hole-electron symmetry. Notably, the effective Hamiltonian H remains unchanged under the transformation d_i to $d_i^\dagger$ and (ϵ_1, ϵ_2) to $(-\epsilon_1, -\epsilon_2)$, highlighting that the supercurrent satisfies $I(-\epsilon_1, -\epsilon_2) = I(\epsilon_1, \epsilon_2)$.

In **Figure 11a** and **b**, quasi-particle current and the total current are plotted against ϕ for various values of thermal bias, respectively. The junction shows a finite (lesser magnitude compared to I_{sc}) thermal response attributed exclusively to the quasi-particle current induced when a small thermal bias (in the linear regime) is applied across the superconducting leads. The magnitude of this thermally induced current increases with the temperature gradient, whereas the supercurrent remains unaffected. The shift in the sinuous-shaped total current I_c in **Figure 11(b)** is only because of the changing magnitude of the QP current with thermal bias across the junction.

6. Conclusion and outlook

We examine the thermoelectric transport properties of an uncorrelated T-shaped DQD Josephson junction using the Green's equation of motion technique. When the coupling of QD_2 is taken into account, the I_{SC} vs. ϕ relationship typically exhibits a sinuous line shape, and the magnitude of the supercurrent diminishes as t_c increases. This suppression occurs because the effective hybrid energy level of both the quantum dots shifts further away from the Fermi level, inhibiting Cooper pair tunneling across the junction within the superconducting gap. In this setup, the side-coupled quantum dot acts as an impurity to the standard SC-QD-SC junction. When the condition $\epsilon_1\epsilon_2 = t_c^2$ is met, a pair of Andreev bound states aligns with the Fermi level, facilitating transport and resulting in a characteristic Fano asymmetric line shape. Specifically, variations in ϵ_1 and t_c can modulate the Fano peak in the I_{sc} vs. ϵ_2 graph, altering the peak position.

The Josephson current follows the relation $I(-\epsilon_1, -\epsilon_2) = I(\epsilon_1, \epsilon_2)$. When a small temperature gradient is introduced across the SC leads, the junction shows a finite but small thermal response primarily due to the current generated by thermally induced quasi-particles outside the SC gap. The magnitude of the thermal current increases with the temperature gradient, while the Josephson current remains unaffected.

The study of ABS' and Josephson current is pivotal for advancements in quantum computing and nanoelectronics. By controlling these elements in DQD-based Josephson junctions, we can improve the efficiency and stability of quantum processing. Josephson junctions are integral components of superconducting electronics, essential for applications in sensing, metrology, and communication. Double quantum dot-based Josephson junctions hold the promise of tunable supercurrents and distinctive noise properties, setting the stage for innovations in future superconducting devices.

When considering the Coulomb interaction U on the quantum dot, obtaining an exact analytical solution becomes challenging, requiring approximations such as Hartree-Fock, Slave-Boson mean-field approximation, or diagrammatic perturbation theory, as well as numerical methods like the Numerical Renormalization Group (NRG) [27–33]. If the intra-dot Coulomb interaction is small relative to other energies of the system, the dot's behavior can be approximated as described in this study. Conversely, strong Coulomb blockade prevents double occupancy, counteracting the induction of superconducting features. Therefore, a thorough understanding of

Andreev bound states (ABSs) necessitates investigating their evolution in relation to various parameters, including gate voltage, intra-dot as well as inter-dot Coulomb interactions, dot-lead tunneling strengths, and the superconducting gap. This remains a highly active research area.

Additionally, this research can be extended to include multi-dot and multi-terminal Josephson junctions, thereby broadening its scope and impact.

Acknowledgements

I acknowledge the financial support from SPARK Internship Program associated with Indian Institute of Technology, Roorkee, Uttarakhand, India, for this study, and I am thankful to Prof. Ajay for giving me the opportunity to work with his research group. I am also grateful to the research scholars Bhupendra Kumar, Sachin Verma, and Veer Pal for their support.

Author details

Aridaman Singh Chauhan[1], Bhupendra Kumar[2*] and Ajay Singh[2]

1 Department of Physics, Indian Institute of Technology, Mandi, Himachal Pradesh, India

2 Department of Physics, Indian Institute of Technology, Roorkee, Uttarakhand, India

*Address all correspondence to: bhupendra_k@ph.iitr.ac.in

References

[1] Franceschi SD, Kouwenhoven LP, Schonenberger C, Wernsdorfer W. Hybrid superconductor– quantum dot devices. Nature Nanotechnology. 2010;**5**:703

[2] Martin-Rodero A, Yeyati AL. Josephson and Andreev transport through quantum dots. Advances in Physics. 2011;**60**:899-958

[3] Meden V. The Anderson–Josephson quantum dot—a theory perspective. Journal of Physics: Condensed Matter. 2011;**31**:163001

[4] Vecino E, Martin-Rodero A, Yeyati AL. Josephson current through a correlated quantum level: Andreev states and π junction behavior. Physical Review B. 2003;**68**:035105

[5] Kouwenhoven LP, Austing DG, Tarucha S. Few-electron quantum dots. Reports on Progress in Physics. 2001;**64**:701

[6] Kastner MA. Artificial atoms. Physics Today. 1993;**46**:24-31

[7] Kumar B, Verma S, Ajay. Phase and thermal-driven transport across T-shaped double quantum dot Josephson junction. Journal of Superconductivity and Novel Magnetism. 2023;**36**:831

[8] Kumar B, Verma S, Chamoli T, Ajay. Josephson transport across T-shaped and seriesconfigured double quantum dots system at infinite- limit. European Physical Journal B. 2023;**96**:168

[9] Kleeorin Y, Meir Y, Giazotto F, Dubi Y. Large tunable thermophase in superconductor – Quantum dot – Superconductor josephson junctions. Scientific Reports. 2016;**6**:35116

[10] Zhu Y, Sun Q-f, Lin T-h. Andreev bound states and the π-junction transition in a superconductor/quantum-dot/ superconductor system. Journal of Physics: Condensed Matter. 2001;**13**:8783

[11] Cheng S-g, Sun Q-f. Josephson current transport through T-shaped double quantum dots. Journal of Physics: Condensed Matter. 2008;**20**:505202

[12] Chamoli T, Ajay. Josephson transport through parallel double coupled quantum dots at infinite-U limit. European Physical Journal B. 2022; **95**:163

[13] Schrieffer JR. Theory of Superconductivity. CRC press; 2018

[14] Timm C. Theory of Superconductivity. Institute of theoretical Physics Dresden; 2012

[15] Bardeen J, Cooper LN, Schrieffer JR. Theory of Superconductivity. Physics Review. 1957;**108**:1175

[16] Tinkham M. Introduction to Superconductivity. Vol. 1. Courier Corporation; 2004

[17] Elzerman JM et al. Quantum dots: A doorway to nanoscale physics. Lecture notes in physics. In: Lecture Notes in Physics. Vol. 667. Berlin, Heidelberg: Springer; 2005

[18] Josephson BD. Possible new effects in superconductive tunnelling. Physics Letters. 1962;**1**:251-253

[19] Jacak L, Hawrylak P, Wojs A. Quantum dots. Springer Science Business Media; 2013

[20] Reed MA. Scientific American Magazine. 1993;**268**:118

[21] Antonio Vettoliere et al. Superconducting Quantum Magnetometer Based on Flux Focusing Effect for High-Sensitivity Applications, Quantum Materials, Devices, and Applications; 2023

[22] Lim SY, Shen W, Gao Z. Carbon quantum dots and their applications. Chemical Society Reviews. 2015;**44**: 362-381

[23] Ekimov AI et al. Absorption and intensity-dependent photoluminescence measurements on CdSe quantum dots: assignment of the first electronic transitions. JOSA. 1993;**B10**:100-107

[24] Pillet J-D et al. Andreev bound states in supercurrent-carrying carbon nanotubes revealed. Nature Physics. 2010;**6**:965-969

[25] Zubarev DN. Double-time Green Functions in Statistical Physics. Soviet Physics Uspekhi. 1960;**3**:320

[26] Feng Q. Study of single impurity Anderson model and dynamical mean field theory based on equation-of-motion method. Diss. Frankfurt (Main), Univ., Diss., 2009. Sov. Phys. Usp. 1960; **3**(320):2009

[27] Rozhkov AV, Arovas DP. Josephson Coupling through a Magnetic Impurity. Physical Review Letters. 1999;**82**:2788

[28] Avishai Y, Golub A, Zaikin AD. Superconductor-quantum dot-superconductor junction in the Kondo regime. Physical Review B. 2003;**67**: 041301

[29] Yeyati AL, Martin-Rodero A, Vecino E. Josephson current in strongly correlated double quantum dots. Physical Review Letters. 2003;**91**:266802

[30] Lim JS, Choi MS. The Andreev states of a superconducting quantum dot: mean field vs exact numerical results. Journal of Physics: Condensed Matter. 2008;**20**: 415225

[31] Karrasch C, Oguri A, Meden V. Josephson current through a single Anderson impurity coupled to BCS leads. Physical Review B. 2008;**77**:024517

[32] T. Yoshioka and Y. Ohashi, Numerical Renormalization group studies on single impurity anderson model in superconductivity, Journal of the Physical Society of Japan. 2000;**69**: 1812-1823

[33] Choi M-S, Lee M, Kang K, Belzig W. Kondo effect and Josephson current through a quantum dot between two superconductors. Physical Review B. 2004;**70**:020502(R)

Chapter 3

Non-Binary Cryogenic Memory Cell Equilibrium Transition Control

Roland Harvey and Zhihua Qu

Abstract

In this chapter, a pulse-based method to facilitate equilibrium transitions in a superconductor memory cell is developed. In a cryogenic environment, non-dissipative storage of non-binary memory states has the potential to drastically increase energy efficiency per operation and decrease the number of memory cells per unit area. Memory cells in consideration are fabricated with Josephson junctions, which are superconducting devices analogous to the classical bit. Dynamics of a memory cell are identical to an array of coupled nonlinear oscillators, which have no known analytic solution. Despite this difficulty, a novel closed-form control technique is developed to design pulses that force stable state transitions in a memory cell. Simulation results are shown to verify the proposed solution.

Keywords: Josephson junctions, coupled oscillators, multi-state transition, analytical design, pulse control

1. Introduction

Classical computing technologies are approaching limits that have not yet been reconciled. Quantum effects begin to take hold with smaller transistors which currently cannot be handled. Concerns of scalability and thermal management in relation to power efficiency and surface area have been sufficiently explored and applied in classical computing for personal computing. However, typical semiconductor-based supercomputers are becoming prohibitively expensive to operate, even with no mention of heat dissipation [1]. Semiconductor technology is ultimately restrictive, but fortunately superconductor research and applications have proven to be a worthwhile alternative. As superconductor technology develops, computing options are diversifying. Cryogenic supercomputing has shown to be more energy efficient and faster than classical counterparts, still with room to improve [2]. In [3, 4], advantages and challenges of superconductor Josephson junction circuits are explained.

Josephson junctions are the building blocks of modern superconducting technologies. Recently, many fabrication techniques and device designs have been investigated which have interesting characteristics. In [5], the authors observe relaxation dynamics in a stochastic bosonic junction, while in [6], the author suggests that gravitational waves can be emitted from the fast oscillations in a hybrid Josephson junction setup. A novel design consisting of inherent orthogonality and a ferromagnetic interface is discussed in [7], which allows phase transitions by changing the direction of magnetization. While such properties are important discoveries, classical physical behavior is investigated for application in the context of cryogenic memory applications.

Cryogenic memory cell fabrication techniques involve Josephson junctions, and researchers have customized cells for efficiency and performance [8]. Architecture and experimentation in cryogenic computing are popular topics [9, 10], however, few approaches have been investigated with regard to scalable memory operation and control. Recently, authors of Ref. [11] presented a computational approach to control memory operations, while authors of Ref. [12] presented an energy-based pulse control design. In Ref. [13], an asymptotic solution for the Josephson junction model was developed, however, accuracy of the approximation depends on parameters in the given system. Applicable analytic options to approach this problem are lacking in number, which is a primary motivator for developing such solutions.

Pulse-based control methods have been useful for various applications including particle accelerator operation as shown in Ref. [14], electromechanical systems in [15], and optics in Ref. [16]. Experimentally, Josephson junctions have been operated by single flux quantum pulses or variations thereof [10, 17].

In the scope of this chapter, a memory cell consisting of an array of three Josephson junctions is operated by an applied pulse. The pulse is injected only in a specified discrete-time interval, while the dynamics of the system behave in continuous-time. The nature of the Josephson junction array is represented by a coupled array of second order nonlinear oscillators, which has no known solution. By combining a designed pulse input based on energy dissipation of undamped oscillators and approximated trajectories from selective linearization, a closed-form control technique is developed that is capable of switching between any stable memory cell equilibria.

2. Problem formulation

Assuming operation in a cryogenic environment, it is well-known that a single Josephson junction obeys the following equation according to the resistively shunted junction (RSJ) model:

$$C\frac{dV}{dt} + \frac{I}{R}V + I_c \sin\phi = v, \tag{1}$$

where the voltage of the junction is defined as

$$V = \frac{\hbar}{2e}\frac{d\phi}{dt}, \tag{2}$$

and where $\hbar$ is Planck's constant, e is electric charge, ϕ is phase, v is the driving current, and C, R, and I_c are the capacitance, resistance, and critical current of the junction, respectively. In dimensionless units, Eq. (1) can be represented as the following:

$$\ddot{\phi} + \gamma\dot{\phi} + \sin\phi = v \tag{3}$$

with $\gamma = \sqrt{\frac{\hbar C}{2eI_c}\frac{1}{RC}}$, which is the same equation as a nonlinear oscillator, or a driven pendulum with a mass of 1 and damping term γ. Just as a pendulum may swing "over the top" and its angle may occupy equilibria at multiples of 2π, the Josephson junction phase behaves the same.

Thus, combining Josephson junctions in superconducting memory cells have the ability to generate and occupy multiple states. A schematic for such a device with three junctions is presented in **Figure 1**.

The memory cell is governed by a set of dynamic equations that behave as coupled versions of the nonlinear oscillator Eq. (3), which is presented as follows [18]:

$$\begin{aligned} \ddot{\phi}_1 + \gamma_1\dot{\phi}_1 + \sin\phi_1 &= v_1 + \mu_1(\phi_2 - \phi_1) \\ \ddot{\phi}_2 + \gamma_2\dot{\phi}_2 + \sin\phi_2 &= v_2 + \mu_1(\phi_1 - \phi_2) + \mu_2(\phi_3 - \phi_2) \\ \ddot{\phi}_3 + \gamma_3\dot{\phi}_3 + \sin\phi_3 &= v_3 + \mu_2(\phi_2 - \phi_3) \end{aligned} \tag{4}$$

where γ_i are the damping coefficients, μ_1, μ_2 are the coupling coefficients based on circuit components, ϕ_i is the angular position (junction phase), and v_i is the driving current input for each junction $i \in \{1,2,3\}$. The structure of v_i consists of an AC pulse $g_i(k_i, t)$ as well as a fixed DC bias current c_i in the form of

$$v_i = g_i(k_i, t) + c_i, \tag{5}$$

where k_i is to be designed. It is assumed that $\int_0^\infty g_i(k_i, t) < \infty$, such that the nature of $g_i(k_i, t)$ is any finite-time pulse which is active in reading or writing time interval $[t_0, t_f]$.

The objective of transition control is to design the pulse amplitude k_i to facilitate state transitions among equilibrium vector $n = [n_1\ n_2\ n_3]^T \in \mathcal{N}$, where $\mathcal{N}$ contains all

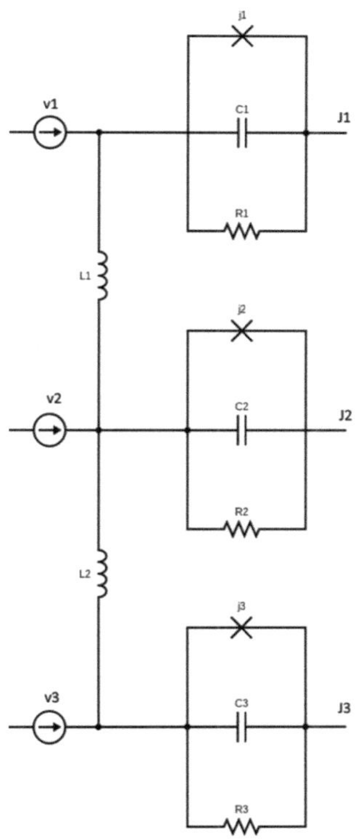

Figure 1.
Memory cell schematic.

possible combinations of memory cell triplets. Inherently, the solution to this control problem is open-loop in the scope of this chapter due to the fact that a feedback control would require continuous reading operations. Computations cannot be done without an external controlling device to handle instructions, similar to that of an arithmetic logic unit (ALU).

3. Nonlinear oscillator analysis

Although the nonlinear dynamics of (4) do not have a closed-form analytical solution, there are certain properties and trajectory characteristics that allow for close approximation of their behaviors. The pendulum-like nature of the dynamics assert that there exist an infinite number of stable and unstable equilibria that are independent from phase velocity and damping. Hence, in the subsequent analysis, undamped companion systems are formed, then through energy analysis the trajectories are corrected by including the work done by damping.

3.1 Equilibrium sets

Equilibrium points are determined from the steady state solution of system (4), which are found by solving the following equations:

$$\begin{aligned} \sin\phi_1 &= c_1 + \mu_1(\phi_2 - \phi_1) \\ \sin\phi_2 &= c_2 + \mu_1(\phi_1 - \phi_2) + \mu_2(\phi_3 - \phi_2) \\ \sin\phi_3 &= c_3 + \mu_2(\phi_2 - \phi_3). \end{aligned} \tag{6}$$

Notice that the individual junction equilibrium points change when neighboring junctions leave the initial equilibrium point.

In **Figure 2**, example partial phase trajectories are shown overlaid with a gradient vector field. Indicated as red asterisks, the equilibrium points for $n_1 = 1, 2$ are shown.

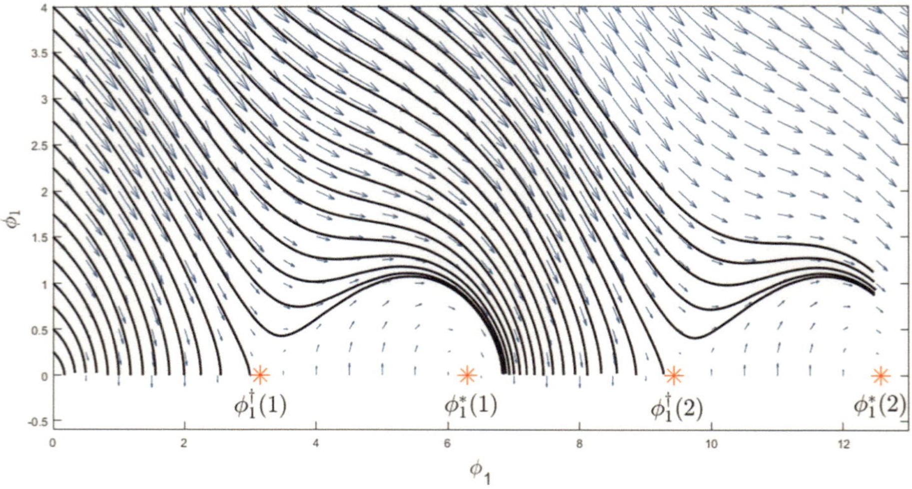

Figure 2.
Isocline trajectories with definitions of equilibria.

Stable equilibrium points are denoted as $\phi_i^*(n_i)$ while unstable equilibrium points are denoted as $\phi_i^\dagger(n_i)$, defined in between $\phi_i^*(n_i)$.

An additional degree of stability must be considered in the context of the RSJ model. As shown in Ref. [18], equilibrium state triplets are not always stable. In order to maintain such unstable states, continuous control is required. Similar to the development in Ref. [19], system (4) is linearized to determine the relationship between equilibrium states and selected bias currents, which is shown in the following expression:

$$\begin{aligned} c_1 &= \mu_1\left(\phi_1^*(n_1) - \phi_2^*(n_2)\right) + \phi_1^*, \\ c_2 &= \mu_1\left(\phi_2^*(n_2) - \phi_1^*(n_1)\right) + \mu_2\left(\phi_2^*(n_2) - \phi_3^*(n_3)\right) + \phi_2^*, \\ c_3 &= \mu_2\left(\phi_3^*(n_3) - \phi_2^*(n_2)\right) + \phi_3^*. \end{aligned} \tag{7}$$

Here, denotes the value of the offset from $2\pi n_i$, while $\phi_i^*(n_i)$ denotes the value of the equilibrium point. It follows that the above systems can be solved as

$$\begin{bmatrix} \phi_1^* \\ \phi_2^* \\ \phi_3^* \end{bmatrix} = \begin{bmatrix} \mu_1 + 1 & -\mu_1 & 0 \\ -\mu_1 & \mu_1 + \mu_2 + 1 & -\mu_2 \\ 0 & -\mu_2 & \mu_2 + 1 \end{bmatrix}^{-1} \times \begin{bmatrix} c_1 - \mu_1(n_1 - n_2) \\ c_2 - \mu_1(n_2 - n_1) - \mu_2(n_2 - n_3) \\ c_3 - \mu_2(n_3 - n_2) \end{bmatrix}^{-1} \tag{8}$$

in which all angles are scaled by 2π. In the scope of this chapter, only stable equilibrium triplets are considered, since discrete pulses are used as the control input.

3.2 Coupled linear oscillator trajectory estimation

Coupling introduces difficulties in phase plane analysis, for instance, phase trajectories are interdependent on neighboring systems. Thus, adjacent junctions are considered as linear systems with no pulse injection. This method will guarantee that while analyzing single junctions, the behavior of phase plane trajectories reflect motion of neighboring junctions that is somewhat deterministic. Fortunately, results for system (4) can be closely approximated by selective linearization of adjacent junctions.

Since dynamics of all junctions are known, the neighbor's behavior can be predicted locally through linearization. The following lemma presents a generalized form of approximating behavior of neighboring junctions in terms of oscillators, since their dynamics are equivalent.

Lemma 1: *The following equation approximately represents the behavior of oscillator i:*

$$\ddot{\phi}_i + \gamma_i \dot{\phi}_i + \sin\phi_i + \mu_i \phi_i - \sum_j \mu_j h_j(\phi_i) = 0, \tag{9}$$

where function $h_j(\phi_i)$ accounts for the behavior of neighboring oscillator j which only depends on the phase angle of oscillator i.

Proof: Consider oscillator j as a neighbor of oscillator i. Treating the dynamics of oscillator j as linear yields the following equation:

$$\ddot{x}_j + \gamma_j \dot{x}_j + \left(1 + \mu_j\right) x_j - \mu_j \phi_i = 0, \tag{10}$$

which has a well known solution:

$$\begin{bmatrix} x_j(t) \\ \dot{x}_j(t) \end{bmatrix} = e^{A_j t} \begin{bmatrix} x_j(0) \\ \dot{x}_j(0) \end{bmatrix} + \int_0^t e^{A_j(t-\tau)} B_j \phi_i(\tau) d\tau \triangleq \begin{bmatrix} h_j(\phi_i) \\ \dot{h}_j(\phi_i) \end{bmatrix}, \tag{11}$$

with $A_j = \begin{bmatrix} 0 & 1 \\ -\left(1+\mu_j\right) & -\gamma_j \end{bmatrix}$, $B_j = \begin{bmatrix} 0 \\ \mu_j \end{bmatrix}$. ■

The nature of solution (11) considers ϕ_i as the only independent variable other than t, which can equivalently be considered as an input. Hence, function $h_j(\phi_i)$ can be calculated at junction i if the dynamics of junction j are known.

3.3 Estimation of dynamics from work

The loss of energy can be determined from the work that would be done by damping, with respect to the force of damping $f_d^i = \gamma_i \dot{\phi}_i(t)$, and the phase angle for junction i as

$$W_i(\phi_i) = \int_0^{2\pi} f_d^i d\phi_i. \tag{12}$$

From Ref. [12], it follows that within once cycle of period T,

$$W_i(t) = \gamma_i \int_0^T \dot{\phi}_i^2 dt. \tag{13}$$

It is also apparent that the damped system's energy is less than the undamped system's energy subject to the work done by damping since the velocity of the undamped system does not decrease while the damping approximation is applied. In summary:

$$E_i^u - W_i \leq E_i, \tag{14}$$

in which W_i may be expressed with respect to time or angle. An analytic expression for $W_i(\phi_i)$ was presented in Ref. [12] based on the elliptic integral, while the integration over time can be easily obtained numerically.

Introduced in [12], the method of using undamped companion systems to predict motion of the damped version is expanded upon based on Eqs. (9) and (12). Since Eq. (12) excludes coupled junctions, the neighboring system approximate behavior of Eq. (9) provides a novel approach to the technique applied in Ref. [12]. The work done including adjacent junction coupling is formed as follows:

$$\hat{W}_i = \int \left[\sum_j \dot{h}_j(\phi_i)^2 + \mu_i \left(h_j(\phi_i) - \hat{\phi}_i \right) \right] dt, \tag{15}$$

where $\hat{\phi}_i$ is an estimation of the phase angle of junction i, which will be determined subsequently. Recall that the square term depends on the linear adjacent systems, which are not subject to pulse injections. This term is included to better model the effect of the motion in junction j caused by junction i, estimated in junction i.

The total energy equation for the undamped version of each junction is

$$\frac{k_i^2 \mathcal{E}_i^2}{\gamma_i^2} = \dot{\varphi}_i^2 + \cos \varphi_i^*(n_i) - \cos \varphi_i, \tag{16}$$

where $\mathcal{E}_i$ denotes the unit energy of the pulse. By equating energies of the damped and adjusted undamped systems and using the work-energy theorem, the estimated companion system is determined from the above and Eq. (15) as

$$\hat{\dot{\phi}}_i = \sqrt{\dot{\varphi}_i^2 + 2\left(\cos \hat{\phi}_i - \cos \varphi_i - W_i - \hat{W}_i\right)}. \tag{17}$$

Eq. (17) serves as a reference trajectory to approximate the behavior of the memory cell. Dynamics of system (17) are simulated in parallel and compared to the actual system dynamics. Trajectories of the real system can be approximated well enough to predict equilibrium state outcomes. The following section provides a method to design pulse gains to achieve various memory cell state transitions.

4. Memory control design

The closed-form control synthesis requires two steps. First, regions of attraction are classified for each stable equilibrium in each junction based on impulse control. Second, pulse gains are designed based on linear objectives determined from impulse regions of attraction. By ensuring the intersection of impulse trajectories with pulse trajectories, a specified equilibrium can be achieved.

4.1 Impulse control

In between control pulses, the energy in the junction dissipates until an equilibrium is reached, thus resulting in the state trajectory behaving as the autonomous system with velocity initial condition $\dot{\phi}_i(t_0)$ determined from the energy in the system after a control pulse, and position initial condition $\phi_i(t_0) < \pi/2 \mod 2\pi$.

Shown in **Figure 3**, sample system trajectories are drawn in solid black, while boundary trajectories are dashed. The blue lines indicate the velocity upper limit for two separate equilibrium points, where the bottom endpoint is an unstable equilibrium point. It can be seen that the phase portrait regions can be linearly approximated to predict which specific equilibrium boundaries should be considered. The aim point is exactly equal to the desired equilibrium point. The slope of isoclines in the phase plane are calculated by evaluating $\dot{\phi}_i(t)/\phi_i(t)$. By extrapolating the initial slope at t_0, which is approximately equal to $-\gamma_i$, the initial velocity required is determined as:

$$\dot{\phi}_i^*(n_i) = \gamma_i 2 n_i \pi \tag{18}$$

which can be rearranged and generalized to form the equation of the red lines:

$$\dot{\phi}_i(t) = -\gamma_i \phi_i(t) + \dot{\phi}_i^*(n_i), \tag{19}$$

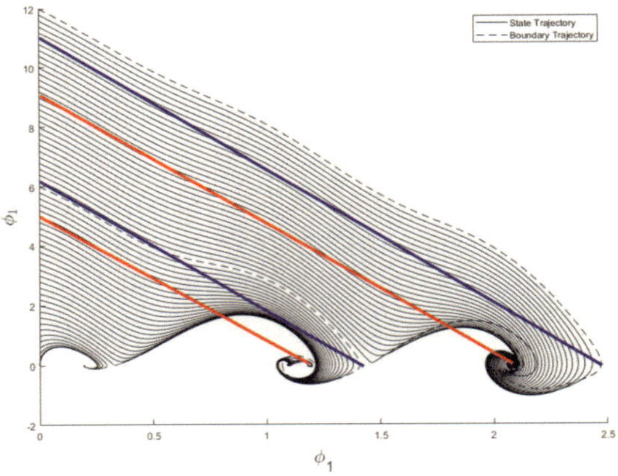

Figure 3.
Equilibrium boundaries.

which is shown for two stable equilibrium points in **Figure 3**. Alternatively using a graphical approach, the following linear equation serves as an objective of the pulse control:

$$\dot{\phi}_i(t) = -\frac{\dot{\phi}_i(t_0)}{\phi_i^*(n_i)}\phi_i(t) + \dot{\phi}_i(t_0), \tag{20}$$

which can be used instead of Eq. (19) if the the result of $\dot{\phi}_i(t_0)/\phi_i^*(n_i)$ does not equal $-\gamma_i$. However, values of $\dot{\phi}_i(t_0)$ must be calculated ahead of time before the pulse.

Although the design is sufficient if impulse control is used, the application in question does not operate under impulses, and thus we require a design of a pulse control to reach the goal of the impulse velocity.

4.2 Pulse control synthesis

Computing operations in real time are generally governed by signals in the form of pulses. A meaningful signal in classical computing is the clock pulse signal, which is usually a square wave, and assists in the scheduling of memory operations. Typically in experimentation, a single flux quantum (SFQ) gaussian pulse is used [20], and recently sinusoidal microwave signals have been explored. For generalization, the amplitude of the pulse, denoted as k_i for the i th junction, will be designed, while the type of pulse signal can be arbitrary.

Theorem 1.1 Consider system (4). To facilitate equilibrium transitions from state $n = \{0,0,0\}$ to state $n = \{n_1, n_2, n_3\}$, the following gain is designed for a pulse injection at all three junctions:

$$\begin{aligned} k_i^2 &= (\gamma_i^2 + 1)\phi_i^2(t_f) - 2\gamma_i\phi_i(t_f)\dot{\phi}_i^*(n_i) + \dot{\phi}_i^*(n_i)^2 \\ &\quad - \begin{bmatrix} x_i(t_0) \\ \dot{x}_i(t_0) \end{bmatrix}^T e^{A_i^T t_f} e^{A_i t_f} \begin{bmatrix} x_i(t_0) \\ \dot{x}_i(t_0) \end{bmatrix}, \end{aligned} \tag{21}$$

where $A \triangleq [A_i | A_j]$ for representative system $\dot{x} = Ax, x \in \mathfrak{R}^n$.

Proof: Given that system (4) behaves linearly in the time interval $t \in [t_0, t_f]$, the following system is representative of the dynamics of a pair of junctions:

$$\begin{aligned} \ddot{x}_i(t) + \gamma_i \dot{x}_i(t) + x_i(t) &= f_i(t) + \mu_i (x_j - x_i), \\ \ddot{x}_j(t) + \gamma_j \dot{x}_j(t) + x_j(t) &= f_j(t) + \mu_j (x_i - x_j), \end{aligned} \tag{22}$$

where $f_i(t), f_j(t)$ are input injections that have a nonzero value in the time interval $[t_0, t_f]$, and zero otherwise, and initial conditions $x_i(t_0), \dot{x}_i(t_0), x_j(t_0), \dot{x}_j(t_0)$ are known. Define the state vector as $x = [x_i\ \dot{x}_i\ x_j\ \dot{x}_j]^T$ such that system (22) can be written as $\dot{x} = Ax$ when $f_i(t) = f_j(t) = 0;\ \ t > t_f$, where

$$A = \left[\begin{array}{cc|cc} 0 & 1 & 0 & 0 \\ -(1+\mu_i) & -\gamma_i & \mu_i & 0 \\ 0 & 0 & 0 & 1 \\ \mu_j & 0 & -\gamma_j & -(1+\mu_j) \end{array}\right] \triangleq [A_i \mid A_j]. \tag{23}$$

The solution to $\dot{x} = Ax$ is $x(t) = e^{At} x(t_0)$ and we can determine that $x(t_f) = e^{At_f} x(t_0)$. If $f(t) \triangleq [f_i(t) f_j(t)]^T \neq 0$, the total energy of system (22) is

$$E(t) = \frac{1}{2} x^T x + \frac{1}{2} \left\| \int_{t_0}^{t_f} f(t) dt \right\|^2. \tag{24}$$

Since pulses are designed before operation, their energy can be determined as a vector of constants: $\int_{t_0}^{t_f} f(t) dt = k$, such that the total energy at the end of the pulse is

$$E(t_f) = \frac{1}{2} \left[e^{At_f} x(t_0) \right]^T e^{At_f} x(t_0) + \frac{1}{2} \|k\|^2. \tag{25}$$

Then, linear Eq. (19) is used as the terminal pulse constraint:

$$\dot{\phi}_i(t_f) = -\gamma_i \phi_i(t_f) + \dot{\phi}_i^{\,*}(n_i). \tag{26}$$

Equating energies (24) and (25) while separating systems i and j yields:

$$\begin{aligned} \dot{\phi}_i^2(t_f) + \phi_i^2(t_f) &= \begin{bmatrix} x_i(t_0) \\ \dot{x}_i(t_0) \end{bmatrix}^T e^{A_i^T t_f} e^{A_i t_f} \begin{bmatrix} x_i(t_0) \\ \dot{x}_i(t_0) \end{bmatrix} + k_i^2, \\ \dot{\phi}_j^2(t_f) + \phi_j^2(t_f) &= \begin{bmatrix} x_j(t_0) \\ \dot{x}_j(t_0) \end{bmatrix}^T e^{A_j^T t_f} e^{A_j t_f} \begin{bmatrix} x_j(t_0) \\ \dot{x}_j(t_0) \end{bmatrix} + k_j^2, \end{aligned} \tag{27}$$

in which (26) can be substituted to result in (21) for i and j when solving for k_i or respectively k_j. ■

The same expression is valid for system j, which can be obtained simply by substituting the index in the above equation. The effect of coupling from the adjacent system is negligible in this time period, which is why μ_j can be ignored for system i, and μ_i can be ignored for system j provided that $\mu_i, \mu_j < 1$. Generally, the velocity

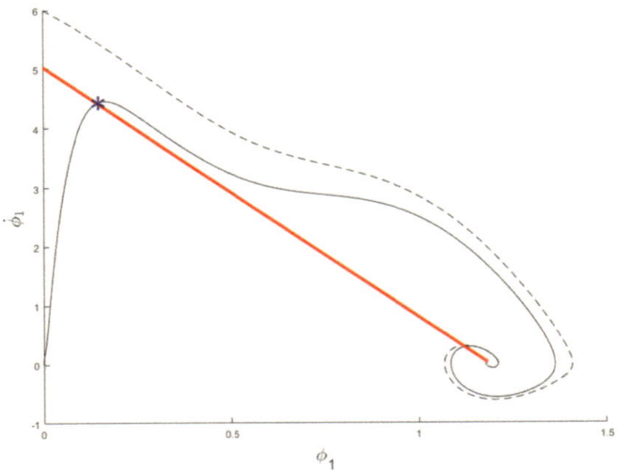

Figure 4.
Example transition with Gaussian pulse.

during the pulse is much greater than the angle traveled. Next, we prescribe $\phi_i(t_f) = \frac{\pi}{2} + \phi_i^*(n_i)$ since we expect the angle to travel no more than $\pi/2$ rads away from the current equilibrium point during the pulse. Now, pulse gain (21) results in a sufficient amount of energy required to reach line (19), and only depends on the current equilibrium point and desired equilibrium point from (18).

As shown in **Figure 4**, the intersection of the red line is close to the maximum point of the trajectory. At that point denoted as a blue asterisk, all energy from the pulse has been transferred to the system such that the system behaves autonomously. Additionally, the maximum point is exactly at time t_f. We see that the angle traveled is no more than $\pi/2$, so our linearization approach for the time interval $[t_0, t_f]$ remains valid.

For further verification, the following section presents simulation results corresponding to specified stable memory states.

5. Simulation

System parameters are prescribed as follows [18, 19]:

$$\begin{aligned} &\gamma_1 = 0.7, \quad \gamma_2 = 1.1, \quad \gamma_3 = 0.7; \\ &c_1 = 1, \quad c_2 = 0.8, \quad c_3 = -1; \\ &\mu_i = 0.1, \quad \forall i. \end{aligned} \tag{28}$$

From the bias current values and coupling coefficients, the initial equilibrium points are determined as:

$$\phi_1^*(0) = (0.1992)2\pi, \qquad \phi_2^*(0) = (0.1187)2\pi, \qquad \phi_3^*(0) = (-.1551)2\pi. \tag{29}$$

In **Figure 5**, a Gaussian pulse is applied to junctions 1 and 2 to achieve the specified transition, which was chosen to exemplify the non-binary capability of the system. The dashed lines represent the trajectories generated from (17), with the values of work done by damping reset every cycle. However, we notice that the same transition can be achieved without applying a pulse to the second junction, as shown in **Figure 6**. The same pulse in junction 1 is used in both simulations of **Figures 5** and **6**.

The corrected undamped trajectory design is scalable enough to design additional transitions, such as the one shown in **Figure 7**. Even though the pulse magnitude is significantly larger, the third junction remains at its zero equilibrium, while the effects of coupling are absorbed into the second junction. Furthermore, since the state {3,1,0} is unstable, the first equilibrium in junction 2 is bypassed. Therefore, certain transitions can be optimized to use less total energy to achieve the same result. Although the reason for such a transition in the case of **Figure 6** can be attributed to the state {2,0,0} being unstable [11]. Even when commanding a transition to an unstable state, it is improbable that the system can settle there when behaving autonomously. Thus, we can assume the reasonable behavior of junction 2 is to be forced out of its current equilibrium due to coupling, and settle at the next equilibrium.

This phenomenon leads to lower energy requirements to reach certain memory cell states if unstable triplet states are known. Furthermore, more options for pulse gains

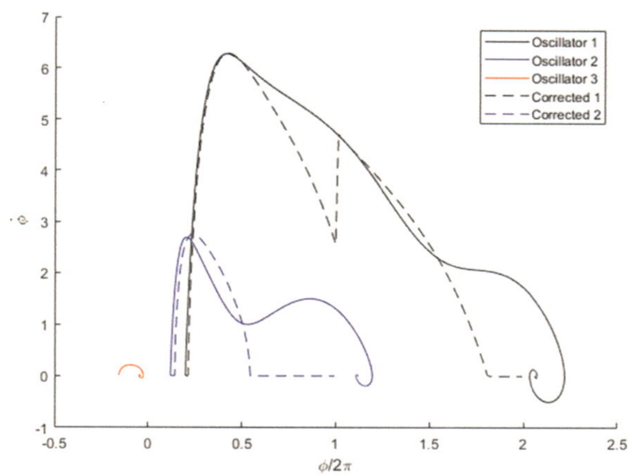

Figure 5.
Three-junction transition {0,0,0} → {2,1,0}.

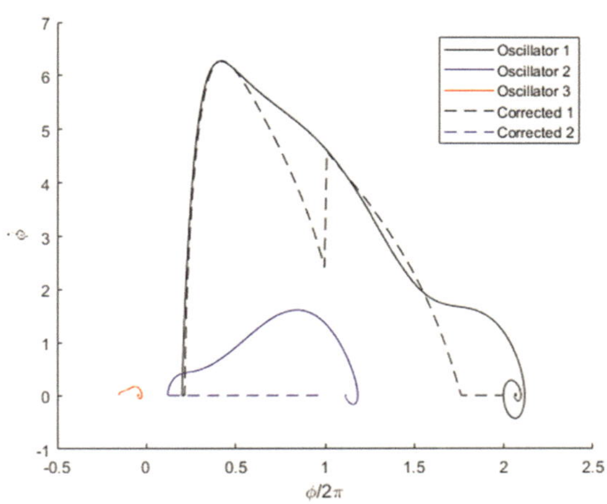

Figure 6.
Three-junction transition {0,0,0} → {2,1,0}, *single pulse.*

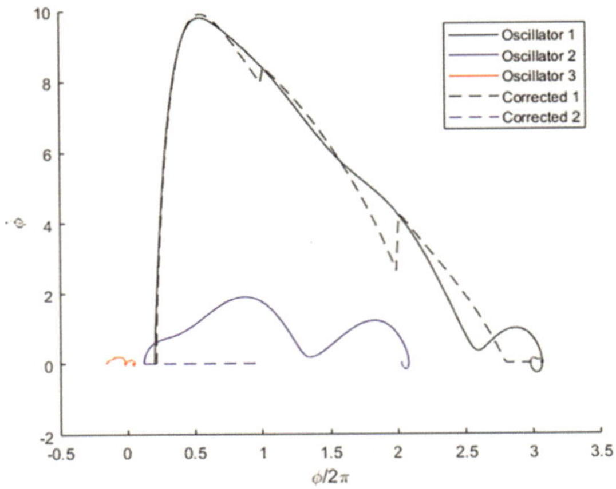

Figure 7.
Three-junction transition {0,0,0} → {3,2,0}, single pulse.

Transition from {0,0,0}	k_1 range
to {1,1,0}	[2.0, 5.2]
to {2,1,0}	[5.3, 11.2]
to {3,2,0}	[11.3, 17.2]

Table 1.
Single pulse transitions.

Transition from {0,0,0}	min k_1	min k_2
to {1,1,0}	0.7	0.6
to {2,1,0}	3.4	3.0
to {3,2,0}	8.7	8.6

Table 2.
Minimum gains for dual pulse transitions.

lie in ranges that are modified by such unstable equilibrium states, which allows for flexibility or larger error margins in practice.

Although it is possible to reach desired equilibria with a single pulse, it is not necessarily more efficient in terms of total cell energy. The ranges of pulse gains for single pulse transitions are shown in **Table 1**, while the minimum gains for dual pulse injection are shown in **Table 2**.

In **Table 1**, the first column indicates the transition being made starting at the {0,0,0} state. The second column presents the range of viable single pulse gains to reach the desired equilibrium triplet for non-minimum transitions.

Table 2 shows the minimum pulse gains required for junctions 1 and 2 corresponding to the applied dual-pulse. In dual-pulse controlled transitions, the energy required based on the gains is less than that of the single pulse transitions except for the case of transition {0,0,0} → {3,2,0}. The dual pulse minimum gains do not inject a lower energy than the single pulse ($8.7^2 + 8.6^2 > 11.3^2$). Thus, we can conclude that at least for smaller equilibrium distances (from 0 to 1 or 0 to 2) the dual pulse is more energy efficient since the energy requirement to reach higher equilibrium points increases exponentially.

6. Conclusion

A pulse-based control design has been developed that facilitates stable memory state transitions in a Josephson junction array. Although the original dynamics have no analytical solution, a close approximation is provided and utilized to assist in the design process. Combining impulse control based on trajectory forecasting and pulse control with linear objectives results in the potential for any state incrementing equilibrium transitions. Gain ranges for Gaussian pulses are classified for selected equilibrium transitions that prove to be effective as well as flexible. Minimum energy gains using dual pulses are determined for more efficient transitions. Other pulse types may be used given that the unit pulse energy is well-defined. Scalability is apparent from the matrix-based design, and is further shown by providing simulation examples with higher energy transitions.

Acknowledgements

Parts of this chapter were previously published in the doctoral thesis by the same author: Roland Harvey. Data-Driven Nonlinear Control Designs for Constrained Systems. 2020. University of Central Florida. Available from: https://stars.library.ucf.edu/cgi/viewcontent.cgi?article=1052&context=etd2020

Author details

Roland Harvey and Zhihua Qu*
Department of Electrical and Computer Engineering, University of Central Florida, Orlando, USA

*Address all correspondence to: qu@ucf.edu

References

[1] Soloviev II, Klenov NV, Bakurskiy SV, Kupriyanov MY, Gudkov AL, Sidorenko AS. Beyond Moore's technologies: Operation principles of a superconductor alternative. Beilstein Journal of Nanotechnology. 2017;**8**(1): 2689-2710

[2] Scott Holmes D, Ripple AL, Manheimer MA. Energy-efficient superconducting computing—Power budgets and requirements. IEEE Transactions on Applied Superconductivity. 2013;**23**(3): 1701610-1701610. DOI: 10.1109/TASC.2013.2244634

[3] Likharev KK. Dynamics of Josephson Junctions and Circuits. London, UK: CRC Press; 2022. ISBN 9781351454193. Available from: https://books.google.com/books?id=7nFbEAAAQBAJ

[4] Likharev KK. Superconductor digital electronics. Physica C: Superconductivity and Its Applications. 2012;**482**:6-18. DOI: 10.1016/j.physc.2012.05.016. ISSN 0921-4534. Available from: https://www.sciencedirect.com/science/article/pii/S0921453412002481

[5] Stefanatos D, Paspalakis E. Relaxation dynamics in a stochastic bosonic Josephson junction. Physics Letters A. 2019;**383**(20):2370-2375

[6] Atanasov V. Gravitational wave emission from quadrupole Josephson junction device. Physics Letters A. 2020; **384**(1):126042

[7] Cheng Q, Zhang K, Ma H. Controllable phase transitions and novel selection rules in Josephson junctions with inherent orthogonality. Physics Letters A. 2018;**382**(9):646-655

[8] Yau J-B, Gibson GW, et al. Hybrid cryogenic memory cells for superconducting computing applications. In: 2017 IEEE International Conference on Rebooting Computing (ICRC), Washington DC, USA. Piscataway, New Jersey, USA: IEEE; 2017. pp. 1-3

[9] Vernik IV, Bolginov VV, Bakurskiy SV, Golubov AA, Kupriyanov MY, Ryazanov VV, et al. Magnetic Josephson junctions with superconducting interlayer for cryogenic memory. IEEE Transactions on Applied Superconductivity. 2013;**23**(3): 1701208-1701208. DOI: 10.1109/TASC.2012.2233270

[10] Volkmann MH, Sahu A, Fourie CJ, Mukhanov OA. Implementation of energy efficient single flux quantum digital circuits with sub-AJ/bit operation. Superconductor Science and Technology. 2012;**26**(1):015002

[11] Glowinski R, López J, Juárez H, Braiman Y. On the controllability of transitions between equilibrium states in small inductively coupled arrays of Josephson junctions: A computational approach. Journal of Computational Physics. 2020;**403**:109023

[12] Harvey R, Zhihua Q. Control of cryogenic memory state transitions in a Josephson junction array. In: 2018 Annual American Control Conference (ACC), Milwaukee, MI, USA. Piscataway, New Jersey, USA: IEEE; 2018. pp. 5671-5676

[13] Liu I, Lee YC. An asymptotic solution for the resistively shunted junction (rsj) model of the Josephson junction. Journal of Low Temperature Physics. 1981;**44**:11-21

[14] Kurimoto Y. A high precision power supply for fast pulse current with a

digital control system. IEEE Transactions on Applied Superconductivity. 2016; **26**(4):1-4

[15] Taylor DG. Pulse-width modulated control of electromechanical systems. IEEE Transactions on Automatic Control. 1992;**37**(4):524-528

[16] Miyamoto D, Mandai K, Kurokawa T, Takeda S, Shioda T, Tsuda H. Waveform-controllable optical pulse generation using an optical pulse synthesizer. IEEE Photonics Technology Letters. 2006;**18**(5):721-723

[17] Nagasawa S, Hinode K, Satoh T, Kitagawa Y, Hidaka M. Design of all-dc-powered high-speed single flux quantum random access memory based on a pipeline structure for memory cell arrays. Superconductor Science and Technology. 2006;**19**(5):S325

[18] Yehuda Braiman N, Nair JR, Imam N. Memory cell operation based on small Josephson junctions arrays. Superconductor Science and Technology. 2016;**29**(12):124003

[19] Braiman Y, Brendan Neschke N, Nair NI, Glowinski R. Memory states in small arrays of Josephson junctions. Physical Review E. 2016;**94**(5):052223

[20] Nair N, Braiman Y. A ternary memory cell using small Josephson junction arrays. Superconductor Science and Technology. 2018;**31**(11):115012

Section 3

Applications of Superconductivity

Chapter 4

Applications of High-Temperature Superconductors in Microwave Devices

Morteza Heidari

Abstract

High-temperature superconductors (HTS) offer significant advancements in the field of microwave technology, particularly in the development of microwave devices and resonators. These superconductors are pivotal in enhancing the performance and efficiency of various medical applications, such as magnetic resonance imaging (MRI) systems, due to their low-loss characteristics and superior conductivity at elevated temperatures. In industrial instruments, HTS are utilized to improve the precision and sensitivity of microwave sensors and communication systems. The integration of HTS in microwave devices and resonators not only optimizes signal quality and reduces energy consumption but also opens new avenues for innovation in both medical and industrial sectors.

Keywords: HTS in medical applications, HTS in industrial instruments, microwave devices, microwave resonators, HTS materials

1. Introduction

High-temperature superconductors (HTS) have revolutionized the fields of material science and engineering, particularly with their significant impact on microwave technology. Discovered in the late 1980s, HTS materials have the remarkable ability to conduct electric current with zero resistance at relatively higher temperatures compared to traditional superconductors, which operate only at cryogenic levels near absolute zero. The practical benefits of HTS materials have led to their adoption in a wide array of applications, from medical imaging systems to telecommunications and defense technologies [1]. The critical temperature at which HTS materials transition into their superconducting state allows them to be integrated into a broader range of technologies, including microwave devices where high performance and minimal signal loss are crucial.

HTS materials, particularly those based on compounds like yttrium barium copper oxide (YBCO), exhibit superconducting properties at temperatures exceeding 77 K, the boiling point of liquid nitrogen. This makes them more cost-effective and practical compared to low-temperature superconductors, which require more complex cooling systems using liquid helium [2].

Figure 1.
A view of a superconductor [6].

Microwave technology, which operates in the frequency range of 300 MHz to 300 GHz, is highly sensitive to energy losses due to resistance in conventional conductors. By incorporating HTS materials into microwave devices, such as resonators, filters, and transmission lines, engineers can significantly reduce energy dissipation, improving overall efficiency. HTS materials exhibit minimal surface resistance at microwave frequencies, which directly enhances the quality factor (Q) of microwave devices. This improvement in Q allows for more precise frequency control, reduced noise, and improved signal clarity in communication systems [3].

In practical applications, HTS devices have been used in satellite communications, radar systems, and even medical imaging devices like MRI machines, where the reduction in energy loss and the enhancement of signal-to-noise ratio (SNR) are critical for high-quality imaging and data transmission [4]. HTS technology also allows for miniaturization of components while maintaining high performance, making it an attractive option for both commercial and industrial microwave applications (**Figure 1**) [5].

2. Key advantages of HTS in comparison to conventional materials

HTS materials offer several advantages over conventional materials, particularly in the realm of microwave devices. Firstly, HTS materials exhibit zero electrical resistance below their critical temperature, which allows for near-lossless transmission of signals, significantly outperforming normal conductors like copper and silver at high frequencies. This zero-resistance property drastically reduces energy loss, thereby increasing the overall efficiency of devices that operate at microwave frequencies [7].

Additionally, HTS materials have a much higher critical current density, meaning they can carry larger amounts of electrical current without losing their superconducting properties. This enables the design of compact and powerful microwave components, which are essential for modern communication systems that demand high performance in smaller, lightweight packages [8]. Moreover, the lower noise characteristics of HTS materials improve the sensitivity and accuracy of devices like microwave sensors and detectors, which are increasingly used in scientific and industrial applications [9].

Another significant advantage is the capability of HTS materials to function in higher magnetic fields, making them suitable for use in environments where conventional superconductors or materials would fail. This makes HTS particularly advantageous for applications such as power generation, transmission, and advanced scientific research [10].

High-temperature superconductors (HTS) operate based on a combination of well-established and emerging scientific theories. These theories provide a foundational understanding of the material properties, mechanisms of superconductivity, and the interaction of superconductors with electromagnetic fields.

2.1 BCS theory (Bardeen-Cooper-Schrieffer theory)

The BCS theory, developed by John Bardeen, Leon Cooper, and Robert Schrieffer in 1957, is a foundational framework for understanding conventional superconductivity. The theory explains how electrons in a metal can form weakly bound pairs, known as Cooper pairs, due to interactions mediated by lattice vibrations (phonons). These pairs move coherently without scattering, resulting in the complete absence of electrical resistance. The BCS theory also introduces the concept of an energy gap, a fundamental characteristic of superconductors, which reflects the energy required to break a Cooper pair. This gap provides insights into the thermal and electromagnetic properties of superconducting materials [11].

Although high-temperature superconductors (HTS) do not strictly follow the BCS framework, the theory's principles provide a basis for exploring superconductivity's quantum mechanical nature. HTS materials, such as cuprates, exhibit behaviors that deviate from conventional superconductors, particularly in their pairing mechanisms and high critical temperatures. Despite these differences, the BCS theory's concepts remain influential in shaping our understanding of superconductivity and guiding the development of alternative theories for HTS.

2.2 Mechanisms of d-wave pairing

High-temperature superconductors differ from conventional superconductors in their pairing symmetry. In HTS materials, the electron pairs are believed to exhibit d-wave symmetry rather than the s-wave symmetry seen in conventional superconductors. This d-wave pairing is characterized by an angular dependence in the energy gap, where the gap has nodes (zeros) along certain directions in momentum space. The pairing mechanism is thought to be driven by antiferromagnetic spin fluctuations rather than phonons, as in the BCS framework. This mechanism is particularly significant in copper-based HTS (cuprates), where the strong electronic correlations and layered crystal structures favor d-wave symmetry [12].

The unique d-wave pairing in HTS explains several experimentally observed phenomena, including anisotropic energy gaps and unconventional responses to magnetic and thermal fields. This mechanism also underpins the higher critical temperatures of HTS compared to conventional superconductors. The study of d-wave pairing continues to be a focus of research, as it holds the key to understanding the fundamental physics of HTS and potentially designing new superconducting materials with even higher transition temperatures.

2.3 Quantum phase transition theory

Quantum phase transitions occur at absolute zero temperature, driven by quantum fluctuations rather than thermal energy. In the context of HTS, these transitions often govern the shift between superconducting and non-superconducting phases under varying external parameters, such as doping levels, pressure, or magnetic fields. Quantum phase transitions are significant in understanding the critical temperatures of HTS and their behaviors near the quantum critical point [13].

This theory is particularly relevant for explaining the complex phase diagrams of HTS materials, where multiple competing phases exist. For example, the interplay between superconductivity, magnetism, and charge-density waves in cuprates can be understood through the lens of quantum phase transitions. Understanding these transitions not only sheds light on the fundamental properties of HTS but also helps in optimizing their performance in practical applications, such as in microwave devices where stability and reliability under extreme conditions are crucial.

2.4 Two-fluid model

The two-fluid model provides a macroscopic description of superconductors, treating them as a mixture of normal electrons and superconducting Cooper pairs. At temperatures below the critical temperature (Tc), the density of superconducting pairs increases while the density of normal electrons decreases, eventually leading to a fully superconducting state. This model is particularly useful for describing the electromagnetic properties of superconductors, such as penetration depth and surface impedance [14].

In HTS, the two-fluid model is essential for understanding their behavior at microwave frequencies. It explains how HTS materials exhibit extremely low surface resistance, which is critical for applications like microwave resonators and filters. The model also provides insights into the frequency-dependent behavior of HTS, enabling the design of high-performance devices with minimal energy dissipation.

2.5 Ginzburg-Landau theory

The Ginzburg-Landau theory offers a macroscopic framework to describe the superconducting state using an order parameter. This parameter reflects the density of Cooper pairs and varies across the superconductor, especially near boundaries or defects. The theory introduces critical fields and coherence lengths, essential for understanding vortex formation and other mixed-state phenomena in HTS [15].

In HTS applications, the Ginzburg-Landau theory plays a pivotal role in designing devices that operate under high magnetic fields. For instance, it helps in optimizing the geometry and material properties of HTS microwave resonators and filters. The theory's ability to describe vortex dynamics is particularly relevant for HTS, where high critical fields and current densities are exploited for advanced technological applications.

2.6 London equations

The London equations, proposed in 1935, provide a phenomenological description of superconductors' electromagnetic properties. They explain two hallmark phenomena: the expulsion of magnetic fields (Meissner effect) and the finite penetration depth of magnetic fields into superconductors. These equations form the basis for understanding the interaction of superconductors with electromagnetic fields at microwave frequencies [16].

For HTS, the London equations are crucial in analyzing their low-loss characteristics in microwave devices. They help in determining the penetration depth, which is typically shorter in HTS than in conventional superconductors, enabling higher efficiency and better performance in resonators and filters. The London equations also underpin the design of HTS devices for communication and sensing, where precise control of electromagnetic properties is required.

3. Electrical and thermal properties of HTS

High-temperature superconductors exhibit unique electrical and thermal properties that distinguish them from conventional conductors and low-temperature superconductors. One of the most significant electrical properties of HTS is their ability to conduct electrical current without resistance once they are cooled below their critical temperature (Tc). This property results in zero power dissipation, making HTS materials ideal for applications requiring efficient energy transmission, such as power grids and magnetic resonance imaging (MRI) systems [17].

The critical current density (Jc), which defines the maximum current that HTS can carry without losing their superconducting state, is another crucial property. HTS materials typically exhibit high Jc values, making them suitable for high-power applications. However, their performance can be limited by magnetic fields, which can suppress superconductivity. Therefore, optimizing HTS materials for use in strong magnetic fields is an ongoing area of research.

Thermally, HTS materials exhibit high thermal conductivity in their normal (non-superconducting) state, but when superconducting, they exhibit a drastic reduction in thermal conductivity due to the absence of electron scattering. This characteristic makes them suitable for applications where heat management is critical, such as in microwave and power transmission systems [18]. However, HTS materials are still sensitive to temperature fluctuations, and maintaining them within a stable cryogenic environment is necessary to ensure optimal performance.

4. HTS in microwave devices

Microwave devices are essential components in a wide array of technologies, including telecommunications, radar systems, medical equipment, and satellite communications. These devices operate in the microwave frequency range.

4.1 Role of HTS in improving device performance

4.1.1 Low loss and high efficiency

One of the most significant advantages of using HTS in microwave devices is the dramatic reduction in energy loss. Conventional conductors like copper experience resistive losses that can severely degrade the performance of microwave components. HTS materials, on the other hand, exhibit zero resistance when cooled below their critical temperature, effectively eliminating resistive losses in microwave circuits.

This property allows HTS-based devices to operate with significantly higher efficiency, making them ideal for applications requiring high power and long-range signal transmission.

For example, in filters and resonators, HTS materials help maintain signal integrity over extended periods, with minimal energy dissipation. This improvement in efficiency translates directly into lower operating costs, reduced cooling requirements, and extended lifetimes for microwave devices [19]. Moreover, the reduction in resistive losses in HTS materials enables the development of smaller and more compact devices, as designers can achieve the same performance with fewer components and less energy.

4.1.2 Enhanced signal-to-noise ratio

Another critical area where HTS materials contribute to improved microwave device performance is the signal-to-noise ratio (SNR). In applications such as radar systems and medical imaging, achieving a high SNR is essential for obtaining clear, accurate data. Resistive losses in conventional materials can introduce noise, which degrades the quality of the signal and reduces the overall performance of the device.

HTS materials provide a solution to this problem by reducing thermal and resistive noise, thereby enhancing the overall SNR of the system. This improved signal clarity is particularly valuable in sensitive applications such as MRI scanners, where clear imaging is necessary for accurate diagnostics. Similarly, in satellite communication systems, the enhanced SNR provided by HTS components allows for more reliable data transmission over long distances, even in noisy environments [20].

4.2 Case studies of HTS microwave devices in research and industry

Several case studies highlight the successful integration of HTS materials into microwave devices across research and industry, demonstrating the material's potential to revolutionize this field.

4.2.1 HTS filters in wireless communication

One of the most notable examples of HTS applications in microwave technology is the development of filters for wireless communication systems. In telecommunications, maintaining clear signal transmission is critical, especially in dense urban environments where multiple devices compete for bandwidth. HTS filters have been implemented in base stations to improve frequency selectivity and reduce interference between channels. These filters, operating with extremely low insertion loss, allow for better channel isolation, resulting in clearer communication and increased system capacity [21].

In a major deployment, an HTS filter system was used in mobile phone base stations in Japan. The deployment showed a marked improvement in call quality and data transmission rates while reducing energy consumption in the base stations. This success highlights the potential for HTS technology to play a significant role in the future of wireless communication infrastructure [22].

4.2.2 HTS resonators in medical imaging

HTS resonators have also been applied in the medical field, particularly in magnetic resonance imaging (MRI) systems. MRI systems rely on high-frequency radio waves to produce detailed images of the human body, and the quality of these images depends largely on the clarity and strength of the transmitted signals. By integrating HTS resonators into MRI systems, researchers have been able to achieve higher image resolution and improved signal clarity. HTS-based coils, for example, reduce noise and improve the quality of the radiofrequency signals, allowing for more accurate diagnostics [23].

HTS significantly enhance the performance of MRI systems, particularly those operating at microwave frequencies. Their integration into critical components like RF and microwave resonators improves signal transmission and reception. HTS resonators exhibit exceptionally low surface resistance at cryogenic temperatures, minimizing signal losses and enabling sharper frequency selectivity. This leads to more precise imaging, especially at microwave frequencies, where conventional materials face higher resistive losses. Another key advantage of HTS in MRI systems is the substantial improvement in signal-to-noise ratio (SNR). The reduced resistive and dielectric losses in HTS components allow for the detection of weaker signals, essential for high-resolution imaging and advanced diagnostic capabilities. Additionally, HTS materials enable the design of compact, efficient MRI coils that generate stronger magnetic fields with lower power consumption. These features make HTS-based coils ideal for portable MRI systems and specialized imaging applications, further advancing the field of medical diagnostics.

In a notable case, HTS resonators were used in experimental MRI systems, where they provided enhanced image resolution without the need for increased signal power or additional cooling infrastructure. The implementation of HTS technology in this field has opened the door for further improvements in medical imaging, offering the potential for faster scans and improved patient outcomes.

4.2.3 HTS in defense and satellite communications

HTS technology has also found applications in defense and satellite communications, where the need for high-performance microwave devices is paramount. In radar systems, for example, HTS materials improve the detection range and accuracy by reducing signal loss and enhancing the SNR. This allows radar systems to detect smaller objects at greater distances, making HTS-based radar systems valuable in both military and civilian air traffic control applications [24].

In satellite communication, HTS components are being used to improve the performance of on-board microwave equipment. HTS-based filters and amplifiers reduce energy consumption and improve data transmission quality, which is crucial for maintaining reliable communication links over long distances in space. These advancements help extend the operational life of satellites and improve the quality of service for satellite-based communication networks.

5. Microwave resonators and filters

5.1 Definition and function of microwave resonators

Microwave resonators are fundamental components in many high-frequency electronic systems, including communication devices, radar systems, and sensors. A resonator is essentially a device that stores electromagnetic energy at a specific frequency, or set of frequencies, for a defined period of time. It functions by allowing certain frequencies (resonant frequencies) to build up inside a cavity or structure while filtering out other frequencies. Resonators play a crucial role in determining the operational bandwidth and frequency stability of microwave systems.

Microwave resonators are used to create high-quality filters and oscillators by selectively amplifying signals within a desired frequency range and suppressing unwanted signals or noise. Their primary characteristics include high quality factor (Q-factor), which defines how efficiently they store energy, and their ability to maintain signal integrity by minimizing loss. In the context of microwave technology, achieving high Q-factors and low losses is essential for optimal device performance, especially in communication and sensing applications [25].

5.2 HTS-based microwave filters for communication systems

Microwave filters are key components in communication systems, as they determine which frequencies are allowed to pass through and which are attenuated. Filters are essential for separating signals in frequency-division multiplexing, reducing interference from other sources, and ensuring clear signal transmission. Traditionally, microwave filters are made from metallic or dielectric materials, but they suffer from resistive losses, especially at higher frequencies, which degrades filter performance.

HTS-based microwave filters represent a major advancement in filter design due to their low resistive losses and high Q-factors. HTS filters, when cooled to cryogenic temperatures, exhibit negligible energy dissipation, making them ideal for applications requiring high frequency selectivity and minimal signal loss. These filters have been successfully integrated into cellular base stations and satellite communication systems, where they improve overall system capacity and performance by providing sharper frequency cutoffs and reducing adjacent channel interference [26].

One notable application of HTS filters is in cellular networks, where they are used to handle high-density traffic and minimize interference between communication channels. By providing higher selectivity than conventional filters, HTS filters allow network operators to accommodate more users in the same frequency band, improving the overall quality of service. These filters also reduce the need for extensive cooling infrastructure, lowering operational costs and extending the life of the equipment [22].

6. HTS in medical applications

6.1 Application of HTS in MRI systems

One of the most prominent applications of high-temperature superconductors (HTS) in the medical field is their use in Magnetic Resonance Imaging (MRI) systems. MRI is a powerful non-invasive diagnostic tool that provides detailed images of the body's internal structures. Traditional MRI systems rely on superconducting magnets, typically made from low-temperature superconductors (LTS), such as niobium-titanium, which require cooling to extremely low temperatures (around 4 K) using liquid helium. However, the introduction of HTS has brought significant improvements to MRI systems, particularly in terms of energy efficiency and imaging quality.

HTS materials, such as Yttrium Barium Copper Oxide (YBCO), offer higher critical temperatures than LTS materials, allowing them to be cooled using liquid nitrogen, which is more readily available and less expensive than liquid helium. This reduces the operational costs and complexity associated with cryogenic cooling in MRI systems. Moreover, HTS coils in MRI systems provide higher quality factors (Q-factors), resulting in enhanced signal-to-noise ratios (SNR) and improved image resolution [27]. The use of HTS technology has enabled the development of smaller, more efficient MRI systems, which are particularly useful for portable or specialized imaging applications.

HTS-based MRI systems also offer advantages in terms of patient comfort and safety. With better imaging quality and reduced noise, HTS-enhanced MRI systems minimize the need for repeated scans, reducing patient exposure to magnetic fields and improving the overall diagnostic process. Additionally, these systems contribute to the miniaturization of MRI technology, paving the way for more compact, accessible MRI units in hospitals and clinics worldwide.

6.2 HTS-based microwave technologies in medical diagnostics and imaging

In addition to MRI systems, HTS-based microwave technologies have found a range of applications in medical diagnostics and imaging. Microwave technologies are commonly used in non-invasive diagnostic procedures, such as breast cancer detection, where microwaves interact with biological tissues to provide valuable diagnostic information. HTS materials have been incorporated into microwave imaging systems to enhance performance by reducing energy loss and improving sensitivity.

One notable example is the use of HTS resonators in microwave-based sensors for medical diagnostics. These resonators exhibit high Q-factors and low insertion losses, allowing for the precise detection of minute changes in biological tissues. HTS-based microwave sensors can be used to detect abnormalities in tissue composition, such as the presence of tumors, by measuring the dielectric properties of the tissue. This has applications in cancer detection, brain imaging, and cardiovascular diagnostics, where high sensitivity and accuracy are crucial [28].

HTS filters and antennas are also being integrated into medical imaging systems to improve their resolution and accuracy. For example, HTS-based antennas are used in microwave imaging techniques, where they help to improve the clarity and depth of images, enabling early detection of diseases. These technologies are especially valuable in situations where conventional imaging techniques, such as X-rays or ultrasound, may be insufficient.

The ability of HTS materials to operate at higher frequencies with minimal losses makes them ideal for medical applications that require precise imaging and diagnostics at microwave frequencies. This is particularly important for imaging soft tissues and detecting early-stage pathologies, where high-resolution, non-invasive techniques are essential.

6.3 Future trends of HTS in medical microwave devices

Looking ahead, the future of HTS in medical applications is promising, with ongoing research and development aimed at further integrating HTS materials into microwave devices for healthcare. One major trend is the miniaturization of HTS-based medical devices, enabling the development of portable and wearable diagnostic tools. For example, HTS-based sensors and imaging devices could be incorporated into wearable health monitors, allowing for continuous, real-time monitoring of patients' physiological parameters.

Another emerging trend is the use of HTS materials in hybrid imaging systems that combine multiple imaging modalities. For instance, HTS-enhanced microwave devices could be used in conjunction with MRI, PET (Positron Emission Tomography), or CT (Computed Tomography) systems to provide more comprehensive diagnostic information. This multimodal approach could significantly improve the accuracy of diagnoses, particularly in complex medical conditions that require detailed anatomical and functional imaging.

As research into HTS materials continues to advance, new applications in areas such as brain-computer interfaces, neural prosthetics, and advanced biosensors are also on the horizon. The unique properties of HTS, such as their ability to conduct large currents without resistance and generate strong magnetic fields, could be leveraged to develop highly sensitive and efficient devices for monitoring neural activity, enabling breakthroughs in the treatment of neurological disorders [29].

Additionally, improvements in HTS fabrication techniques and cryogenic cooling technologies are expected to drive down the costs of HTS-based medical devices, making them more widely accessible to healthcare providers. This will likely lead to the broader adoption of HTS technologies in both developed and developing countries, improving the overall quality of medical care and diagnostic capabilities worldwide.

7. HTS in industrial and commercial instruments

7.1 HTS in microwave sensors and detectors

High-temperature superconductors (HTS) have become increasingly vital in the development of advanced microwave sensors and detectors for industrial and commercial applications. HTS-based microwave sensors can detect extremely small changes in electromagnetic signals, making them ideal for use in precision measurement tools. These sensors are used in applications that require high accuracy and minimal noise, such as detecting material defects in manufacturing processes or monitoring the electromagnetic properties of the atmosphere. For example, HTS microwave detectors have been employed in aerospace industries for the precise detection of temperature fluctuations in the atmosphere, which is crucial for satellite and aircraft safety [30].

Furthermore, HTS microwave detectors are used in imaging systems to identify structural weaknesses in materials, particularly in industries like aerospace, automotive, and construction, where the integrity of materials is crucial. These sensors provide enhanced detection capabilities, allowing for earlier identification of potential faults or wear, reducing downtime, and improving safety standards.

7.2 Use of HTS in wireless communication and satellite technologies

HTS-based devices have been implemented in ground-based satellite communication systems, where they help to optimize signal reception and data processing. Their ability to maintain high performance at cryogenic temperatures makes HTS devices particularly suitable for space applications, where low temperatures are prevalent. These systems contribute to more efficient satellite communication networks, providing clearer, more reliable data transmission, which is crucial for both commercial and military satellites [31].

7.3 Industrial applications: Precision measurement tools and microwave instruments

One of the key advantages of HTS materials in these applications is their ability to operate with minimal energy loss, even at high frequencies. This allows HTS-based devices to maintain high precision and stability, which is essential in fields like metrology, material testing, and process control.

HTS-based microwave instruments are used in various industrial applications for precise measurements of electromagnetic properties, such as dielectric constants, permeability, and conductivity. These measurements are crucial in industries such as semiconductor manufacturing, where material properties must be closely monitored to ensure product quality. HTS sensors and resonators are also employed in non-contact measurements, providing highly accurate results without the need for physical interaction with the material being tested.

Another notable industrial application of HTS is in high-frequency communication instruments used in remote sensing and geophysical surveys. HTS-enhanced microwave instruments are capable of detecting and analyzing electromagnetic waves reflected from the Earth's surface, providing valuable data for industries such as oil and gas exploration, mining, and environmental monitoring. These instruments offer improved sensitivity and range compared to traditional tools, enabling more accurate detection of underground resources and environmental changes.

In commercial settings, HTS devices are used in microwave imaging systems for quality control and security screening. HTS-based systems have the ability to detect fine details in materials, making them ideal for inspecting products on assembly lines or screening luggage at airports. Their low-loss, high-sensitivity properties allow for

Figure 2.
MIT laboratory of high-temperature superconductors [32].

more detailed and efficient scanning, reducing the likelihood of defects or threats being missed during inspections.

HTS technology is also making its way into the consumer electronics market, particularly in applications like wireless charging, where efficient energy transfer is critical. HTS-based microwave circuits have the potential to increase the efficiency of wireless charging systems, providing faster charging times and reducing energy waste in devices ranging from smartphones to electric vehicles (**Figure 2**) [22].

8. Challenges and limitations

8.1 Technical and manufacturing challenges of HTS integration

The integration of high-temperature superconductors (HTS) into microwave devices and other applications presents several technical and manufacturing challenges. While HTS materials such as Yttrium Barium Copper Oxide (YBCO) offer significant advantages in terms of low energy loss and high performance, they are difficult to produce and integrate into practical devices. One of the key challenges is the fabrication of HTS films with high quality and uniformity. The process of growing thin films of HTS materials requires precise control over temperature, pressure, and deposition techniques to ensure the desired superconducting properties. Any deviation in the fabrication process can lead to defects that degrade the performance of the HTS devices, particularly in terms of critical current density and overall efficiency [33].

Another significant technical challenge is the brittleness of HTS materials, which makes them more difficult to work with compared to conventional materials like copper. This brittleness limits the flexibility and durability of HTS devices, especially when they need to be shaped into complex geometries for certain microwave components, such as resonators and filters. Additionally, the integration of HTS materials with other components, such as dielectric substrates and metallic connectors, requires advanced techniques to ensure proper thermal and electrical contact without compromising the superconducting properties of the HTS.

The limited availability of suitable substrates for growing HTS films is another constraint. For optimal performance, HTS films need to be deposited on substrates that match their crystal structure and thermal expansion properties. Currently, substrates like lanthanum aluminate (LAO) and magnesium oxide (MgO) are commonly used, but they are expensive and difficult to scale up for industrial production. Finding cost-effective and scalable materials for HTS film growth remains an ongoing challenge in the field [34].

8.2 Limitations in cooling systems for HTS devices

HTS materials operate at higher temperatures than low-temperature superconductors (LTS), but they still require cryogenic cooling systems to maintain their superconducting state. Typically, HTS materials need to be cooled to around 77 K using liquid nitrogen, which is significantly more manageable than the 4 K required for LTS materials using liquid helium. However, cooling remains one of the major limitations of HTS devices, especially when scaling up for industrial applications.

The reliance on cryogenic cooling systems adds complexity and cost to HTS-based devices, limiting their widespread adoption in certain industries. These systems require regular maintenance and are not always feasible in environments where access to cryogenic infrastructure is limited. For example, while HTS microwave filters could theoretically be used in telecommunications infrastructure, the need for constant cooling has hindered their deployment in large-scale communication networks [34].

Furthermore, the efficiency and reliability of cryogenic cooling systems are not always ideal. Maintaining stable and uniform cooling over long periods can be challenging, particularly in environments with fluctuating temperatures. Any significant deviation in temperature can cause the HTS material to lose its superconducting properties, resulting in device failure or degraded performance. This sensitivity to temperature fluctuations limits the operational flexibility of HTS devices and restricts their use in certain applications.

Efforts to develop more efficient and compact cryocooling systems are underway, but they are still in the research and development phase. Until these cooling technologies are optimized for HTS applications, the reliance on cryogenics will remain a key limitation of HTS devices in both industrial and commercial settings.

8.3 Cost and scalability issues

One of the most significant barriers to the widespread adoption of HTS technology is the high cost of both the materials and the associated infrastructure. HTS materials, particularly high-quality YBCO films, are expensive to produce due to the complex fabrication processes involved. The need for specialized equipment, high-purity chemicals, and precision-controlled environments drives up the manufacturing costs. Additionally, the cryogenic cooling systems required to maintain HTS operation contribute to the overall expense, particularly when factoring in the cost of liquid nitrogen or other cooling agents, as well as the energy required to maintain cryogenic temperatures.

The scalability of HTS technology is another concern. While HTS-based devices have shown great promise in research and specialized applications, scaling these technologies for mass production is challenging. The fabrication of large-scale HTS devices requires consistent quality over extended areas of HTS films, which is difficult to achieve with current deposition techniques. This has restricted the commercialization of HTS devices to niche markets, such as MRI systems and satellite communications, where performance demands justify the high costs [35].

Moreover, the global supply chain for HTS materials and the specialized equipment needed to produce them is limited. As a result, economies of scale have not yet been realized, further driving up the costs of HTS technologies. For widespread adoption, breakthroughs in both material science and manufacturing processes are required to reduce production costs and improve scalability.

In addition to material and manufacturing costs, the deployment of HTS devices in industrial settings often involves retrofitting or upgrading existing infrastructure to accommodate cryogenic cooling systems and specialized components. This adds further expense and logistical complexity, making it less attractive for industries that operate on tight margins.

9. Future prospects and innovations

9.1 Emerging trends in HTS materials and device technology

The future of high-temperature superconductors (HTS) is promising, driven by ongoing advancements in material science and device technology. One of the most notable emerging trends in HTS research is the exploration of new materials that can achieve superconductivity at even higher temperatures. Recent studies on iron-based superconductors and hydrides have shown potential for achieving superconductivity at temperatures closer to room temperature, albeit under extreme pressure conditions. These developments could eventually lead to the discovery of new HTS materials that can operate at significantly higher temperatures, reducing the reliance on expensive and complex cryogenic systems [36].

In addition to new materials, improvements in the fabrication and processing of HTS thin films are enhancing the performance and scalability of HTS devices. Techniques such as pulsed laser deposition (PLD) and chemical vapor deposition (CVD) are being refined to produce films with greater uniformity and fewer defects, leading to higher critical current densities and improved microwave performance. Furthermore, research into the integration of HTS with other advanced materials, such as graphene and topological insulators, is opening new avenues for the development of hybrid devices with enhanced properties [37].

The miniaturization of HTS devices is another important trend. As researchers develop more efficient and compact HTS components, the potential for integrating HTS into everyday technologies, such as consumer electronics and mobile communication systems, is growing. This miniaturization could make HTS technology more accessible and cost-effective for a wider range of applications, from portable medical devices to advanced wireless communication systems [38].

9.2 Potential advancements in HTS microwave applications

The application of HTS in microwave technology is set to expand significantly as new innovations are realized. In the near future, HTS materials are expected to play an increasingly important role in advanced communication systems, including 5G and 6G networks. HTS-based microwave filters and resonators, with their low loss and high-quality factors, are well-suited for managing the high-frequency signals required by next-generation communication technologies. These devices can offer enhanced frequency selectivity, reducing interference and improving the overall capacity and reliability of communication networks.

In the realm of satellite communication, HTS materials are likely to be used in the next generation of satellite payloads. The superior performance of HTS components in low-temperature space environments makes them ideal for improving the signal strength, bandwidth, and efficiency of satellite communication systems. As the demand for high-speed global connectivity grows, HTS technology will be pivotal in delivering the required advancements in satellite communication infrastructure.

Moreover, the use of HTS in quantum computing and quantum communication is an exciting area of future development. HTS materials can be used to create high-efficiency microwave resonators and detectors for quantum systems, helping to bridge the gap between classical and quantum technologies. These applications have the potential to lead to breakthroughs in computational power and secure communication systems, with significant implications for industries ranging from finance to national security.

9.3 Outlook on commercial adoption of HTS in the microwave industry

The commercial adoption of HTS technology in the microwave industry is poised for growth, driven by the increasing demand for high-performance, energy-efficient devices. One of the key factors that will determine the pace of commercial adoption is the ability to reduce the cost and complexity of HTS systems. Advances in material synthesis, fabrication techniques, and cryogenic cooling technologies will be crucial in making HTS devices more affordable and easier to integrate into existing infrastructure.

In the telecommunications industry, HTS-based microwave filters and resonators are already being used in some high-performance applications, particularly in areas where signal clarity and bandwidth are critical. As the cost of HTS components decreases and more efficient cooling solutions are developed, it is expected that HTS technology will become more widespread in cellular base stations, satellite communication systems, and other high-frequency applications.

The medical sector is another area where HTS technology is likely to see significant commercial adoption. HTS-based MRI systems and microwave diagnostic tools offer superior performance compared to traditional technologies, and as these systems become more affordable, their use in hospitals and clinics is expected to grow. The ability of HTS devices to operate at higher frequencies with minimal energy loss makes them ideal for medical applications that require precision and reliability.

However, challenges related to cost and scalability remain significant barriers to widespread commercial adoption. The development of cost-effective cooling systems, such as cryocoolers that do not require constant maintenance or the replacement of cryogenic fluids, will be essential in making HTS technology more commercially viable. In addition, the scalability of HTS manufacturing processes must improve to meet the demands of mass production in industries like telecommunications and consumer electronics.

10. Conclusion

High-temperature superconductors (HTS) have emerged as a transformative technology in the field of microwave devices, providing significant improvements in performance, efficiency, and signal quality. The unique properties of HTS, such as near-zero electrical resistance and exceptional current-carrying capacity at elevated temperatures, have allowed for the development of devices with remarkably low energy dissipation, enhanced signal-to-noise ratios, and improved frequency selectivity. These advantages have made HTS indispensable in a variety of applications, from advanced communication systems to high-precision medical imaging technologies.

The integration of HTS into microwave devices has already led to substantial advancements. In communication systems, HTS-based filters and resonators offer unprecedented performance in terms of bandwidth, signal clarity, and interference reduction. Similarly, in the medical field, HTS has enabled the development of more accurate and efficient MRI systems, as well as new microwave-based diagnostic tools. These contributions highlight the significant impact of HTS on critical technologies, demonstrating the potential of superconductors to revolutionize entire industries [39].

Despite the impressive progress, the full potential of HTS in microwave technology is yet to be realized. As ongoing research addresses challenges such as cooling requirements, manufacturing scalability, and material costs, the next generation of HTS devices is likely to unlock new possibilities. The future of HTS in microwave applications promises further improvements in device miniaturization, power efficiency, and functionality. Emerging technologies, such as 5G and 6G communication systems, quantum computing, and advanced sensing applications, stand to benefit greatly from HTS developments [40].

Overall, HTS has already established its value in microwave devices, offering unique solutions to performance challenges that conventional materials cannot match. As research continues to push the boundaries of HTS materials and device integration, we can expect further breakthroughs that will shape the future of microwave technology across various sectors, including telecommunications, healthcare, and industrial instrumentation. HTS remains poised to drive innovation and enhance the capabilities of microwave systems for years to come.

Author details

Morteza Heidari
Shahed University, Tehran, Iran

*Address all correspondence to: morteza.heidary@shahed.ac.ir

References

[1] Bednorz JG, Müller KA. Possible high Tc superconductivity in the Ba-La-Cu-O system. Zeitschrift fur Physik B: Condensed Matter. 1986;**64**:189-193. DOI: 10.1007/BF01303701

[2] Poole CP, Farach HA, Creswick RJ, Prozorov R. Superconductivity. 2nd ed. Academic Press; 2007. 670 p

[3] Chaloupka HJ. Applications of Superconductivity. 2nd ed. Vol. 365. Dordrecht, Netherlands: Springer; 2000. pp. 295-383. DOI: 10.1007/978-94-017-0752-7_6

[4] Weinstock H. Applications of Superconductivity. 1st ed. Dordrecht, Netherlands: Springer; 1996. 692 p. DOI: 10.1007/978-94-017-0752-7

[5] Lancaster MJ. Passive Microwave Device Applications of High-Temperature Superconductors. 1st ed. New York, United States of America: Cambridge University Press; 1997. 360 p. DOI: 10.1017/CBO9780511526688

[6] Available from: https://www.esquire.com/es/ciencia/a62411401/griffiths-singularidad/

[7] Richards PL. High Tc superconducting detectors. Proceedings of the IEEE. 1994;**82**:595-606

[8] Lancaster MJ, Chiang JY. Superconducting microwave filters. IEEE Transactions on Applied Superconductivity. 1994;**4**:100-106

[9] Collings EW, Sumption MD. High-temperature superconducting materials. Annual Review of Materials Research. 2001;**31**:283-329

[10] Orlando TP, Delin KA. Foundations of Applied Superconductivity. 1st ed. United Kingdom: Addison-Wesley; 1991. 584 p. DOI: 10.1063/1.2810145

[11] Bardeen J, Cooper LN, Schrieffer JR. Theory of superconductivity. Physics Review. 1957;**108**(5):1175-1204

[12] Scalapino DJ. The case for d-wave pairing in the cuprate superconductors. Physics Reports. 1995;**250**(6):329-365

[13] Sachdev S. Quantum phase transitions. Reviews of Modern Physics. 2003;**75**(4):913-932

[14] Hirsch JE. Two-fluid theory for superconductors. Physical Review B. 1985;**31**(7):4403-4412

[15] Ginzburg VL, Landau LD. On the theory of superconductivity. Zhurnal Eksperimental'noi i Teoreticheskoi Fiziki. 1950;**20**:1064-1082

[16] London F, London H. The electromagnetic equations of the supraconductor. Proceedings of the Royal Society of London A. 1935;**149**(866):71-88

[17] Tinkham M. Introduction to Superconductivity. 2nd ed. United States of America: McGraw-Hill; 1996. 480 p

[18] Gurevich A. Challenges and opportunities for applications of unconventional superconductors. Annual Review of Condensed Matter Physics. 2014;**5**:35-56. DOI: 10.1146/annurev-conmatphys-031113-133822

[19] Pinto R et al. Microwave surface resistance of superconducting $YBa_2Cu_3O_{7-\delta}$ thin films. IETE Journal of Research. 2015;**45**(3-4):229-232. DOI: 10.1080/03772063.1999.11416101

[20] Wollman DA, et al. Experimental determination of the symmetry of the superconducting pairing state in YBCO [PhD thesis]. United States of America: University of Illinois; 1996. 157 p

[21] Zou T et al. HTS filter subsystem for future mobile communication system. Science in China Series F: Information Sciences. 2008;**51**:1384-1390. DOI: 10.1007/s11432-008-0085-2

[22] Knack A. Design and implementation of HTS technology for cellular base stations: An investigation into improving cellular communication [PhD thesis]. New Zealand and Australia: Massey University and James Cook University; 2006. 172 p

[23] Lvovsky Y et al. Novel technologies and configurations of superconducting magnets for MRI. Superconductor Science and Technology. 2013;**26**:093001. DOI: 10.1088/0953-2048/26/9/093001

[24] Zhang T et al. High-Tc superconducting microwave millimeter devices and circuits—An overview. IEEE Journal of Microwaves. 2022;**2**:374-388. DOI: 10.1109/JMW.2022.3171675

[25] Craven GF, Mok CK. The design of evanescent mode waveguide bandpass filters for a prescribed insertion loss characteristic. IEEE Transactions on Microwave Theory and Techniques. 1971;**19**:295-308. DOI: 10.1109/TMTT.1971.1127503

[26] Paramentier RD, Pedersen NF. Nonlinear Superconducting Devices and High-Tc Materials. 1st ed. Singapore: World Scientific; 1994. 478 p

[27] Labbe A et al. Recent advances and challenges in the development of radiofrequency HTS coil for MRI. Frontiers in Physics. 2021;**9**:1-13. DOI: 10.3389/fphy.2021.705438

[28] Velluire-Pellat Z et al. Hybrid quantum systems with high-Tc superconducting resonators. Scientific Reports. 2023;**13**:1-10. DOI: 10.1038/s41598-023-41472-z

[29] Johns LS et al. High Temperature Superconductivity in Perspective. Washington, United States of America: US Congress Office Tech Assessment; 1990. 129 p

[30] Newman N. High-temperature superconducting microwave devices: Fundamental issues in materials, physics, and engineering. Journal of Superconductivity. 1993;**6**:119-160. DOI: 10.1007/BF0062574

[31] Hein MA. Potential of high-temperature superconductors for short-term microwave application. High-Temperature Superconductivity. 1996;**53**(135):197

[32] Available from: https://news.mit.edu/2024/tests-show-high-temperature-superconducting-magnets-fusion-ready-0304

[33] Shimoyama JI, Motoki T. Current status of high temperature superconducting materials and their various applications. IEEJ Transactions on Electrical and Electronic Engineering. 2024;**19**:292-304. DOI: 10.1002/tee.23976

[34] MacManus-Driscoll JL, Wimbush SC. Processing and application of high temperature superconducting coated conductors. Nature Reviews Materials. 2021;**6**:587-604. DOI: 10.1038/s41578-021-00290-3

[35] Yao C, Ma Y. Superconducting materials: Challenges and opportunities for large-scale applications. iScience. 2021;**24**:1-17. DOI: 10.1016/j.isci.2021.102541

[36] Wu H. Recent development in high temperature superconductor: Principle, materials, and applications. Journal of Applied and Computational Engineering. 2024;**63**:153-171. DOI: 10.54254/2755-2721/63/20241015

[37] Sibanda D et al. A mini review on thin film superconductors. Processes. 2022;**10**:1-40. DOI: 10.3390/pr10061184

[38] Yangbo B et al. A miniaturized HTS microwave receiver front-end subsystem for radar and communication applications. Physica C: Superconductivity and its Applications. 2010;**470**:617-621. DOI: 10.1016/j.physc.2010.06.001

[39] Coombs TA et al. High-temperature superconductors and their large-scale applications. Nature Reviews Electrical Engineering. 2024;**1**:788-801. DOI: 10.1038/s44287-024-00112-y

[40] Anlage SM. Microwave superconductivity. IEEE Journal of Microwaves. 2021;**1**:389-402. DOI: 10.1109/JMW.2020.3033156

Chapter 5

Superconducting Devices: From Quantum Computing to Energy Transmission

Belqees Hassan

Abstract

Superconducting devices, leveraging the unique properties of zero resistance and the Meissner effect, are transforming diverse technological fields. This chapter explores their applications, from quantum computing to energy transmission and medical imaging. Superconducting quantum computers, employing superconducting qubits and circuits, promise breakthroughs in computing power and speed due to longer coherence times, higher fidelity, and reduced noise. Superconductors revolutionize energy transmission by enabling lossless energy transfer through high-current carrying cables, thus enhancing grid efficiency and reducing energy waste. Additionally, superconducting magnetic energy storage (SMES) systems offer efficient and rapid energy storage for grid stabilization and renewable energy integration. Superconducting technology significantly enhances medical imaging, particularly Magnetic Resonance Imaging (MRI), by enabling the generation of powerful magnetic fields, leading to clearer images and faster scan times. Despite challenges in cost and material limitations, ongoing research aims to overcome these hurdles, paving the way for wider adoption of these transformative technologies.

Keywords: superconducting devices, quantum computers, qubits, superconducting magnetic energy storage, magnetic resonance imaging

1. Introduction

The journey of superconductivity began in 1911 with Dutch physicist Heike Kamerlingh Onnes's groundbreaking discovery of the phenomenon in mercury at extremely low temperatures. This marked the beginning of a revolution in understanding the behavior of materials at low temperatures. The discovery paved the way for the development of various superconducting materials, each with unique properties and applications. Superconducting materials exhibit zero electrical resistance and perfect diamagnetism below a critical temperature [1–6]. This exceptional property enables lossless energy transfer, making superconductors highly desirable for energy transmission and storage applications. This chapter delves into superconducting devices, highlighting their distinctive properties and applications across different technological fields. It begins by exploring superconducting qubits and their pivotal role in advancing quantum computing capabilities. Superconductors offer benefits such as longer coherence times, higher fidelity, and reduced noise in quantum computing applications. The chapter also discusses the role of superconductors in energy transmission and storage, including their utilization in power transmission lines and superconducting magnetic energy storage (SMES) systems for efficient and rapid energy storage solutions. Furthermore, the chapter explores how superconducting magnets, known for their ability to generate powerful magnetic fields, are revolutionizing medical imaging, particularly in MRI technology. Superconducting MRI systems offer advantages, such as enhanced image clarity, faster scan speeds, and improved diagnostic accuracy. General, this exploration aims to display the diverse applications of superconductors while shedding light on the challenges and opportunities that lie ahead in this rapidly evolving field.

2. Applications of superconducting qubits in quantum computing

Superconducting qubits are at the forefront of the quantum computing race due to their ability to harness the unique properties of superconductors to create and manipulate quantum bits (qubits), which are the fundamental units of quantum information processing [7, 8]. These qubits are essential for performing quantum computations and enabling the development of practical quantum computers. There are different types of superconducting qubits used in quantum computing, such as transmon qubits and flux qubits [9, 10]. Transmon qubits consist of a superconducting loop interrupted by a Josephson junction, while flux qubits utilize a superconducting loop interrupted by two Josephson junctions to encode information in the direction of persistent current flowing through the loop. Control of superconducting qubits is achieved through the application of microwave pulses to the qubit's resonant frequency [11, 12]. By precisely adjusting the parameters of these pulses, researchers can manipulate the qubit's quantum state, transitioning it between the states |0 > and |1 > or creating superpositions. To measure the state of a qubit, special measurement circuits are used, which interact with the qubit to reveal its state by monitoring changes in current or voltage within the superconducting circuit [13, 14]. **Figure 1** illustrates the various components involved in the control and measurement of

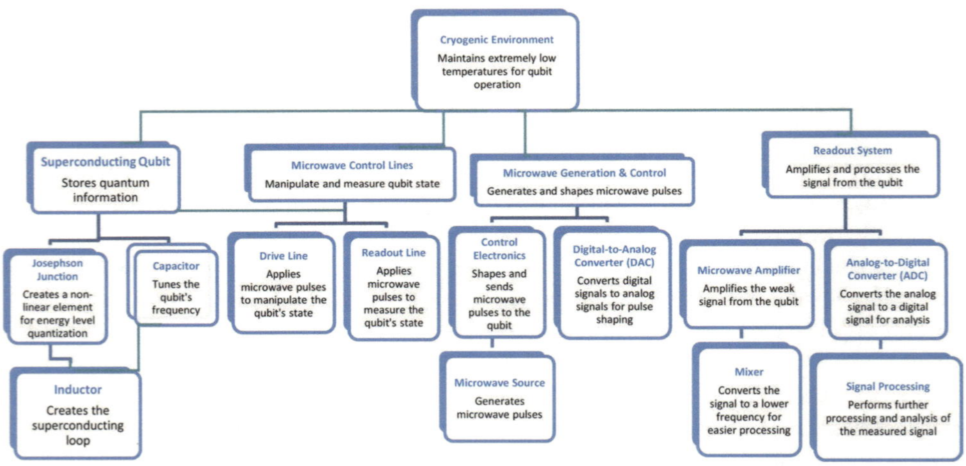

Figure 1.
Superconducting qubit control system components.

superconducting qubits. As shown in the diagram, microwave pulses are generated and controlled using specialized electronics. These pulses are then applied to the qubit through a microwave transmission line, manipulating the qubit's energy levels and inducing transitions between its quantum states. The readout system measures the qubit's state by detecting the changes in current or voltage within the superconducting circuit.

Superconducting qubits can be entangled, a crucial feature of quantum computing that allows for the creation of correlated quantum states between multiple qubits. This entanglement is essential for performing powerful computational operations using techniques like controlled quantum gates (CNOTs) [15]. These qubits are also used to implement various quantum algorithms, such as Shor's algorithm for factoring numbers, Grover's algorithm for database searching, and quantum simulations for exploring complex systems [16]. Superconducting qubits offer several advantages, including long coherence times, high fidelity in quantum operations, and scalability for integration into large-scale circuits. However, they also pose challenges, such as the need for cryogenic cooling systems to maintain extremely low temperatures for operation and the engineering complexities of achieving efficient qubit coupling [17–19]. Superconducting Quantum Interference Devices (SQUIDs) represent advanced magnetometers that leverage the unique properties of superconductors and Josephson junctions to detect extremely faint magnetic fields [20, 21]. The operation of SQUIDs is based on intricate principles that involve the Josephson Effect and the quantum tunneling phenomenon through superconducting junctions. The Josephson Effect in SQUIDs relies on the behavior of superconducting materials separated by a thin insulating barrier, where electrons can tunnel through, generating a supercurrent [22].

This tunneling process is a quantum mechanical effect, and the supercurrent is influenced by the phase difference between the superconducting wave functions on either side of the junction, which is sensitive to magnetic fields [23]. A typical SQUID structure consists of a superconducting loop with two Josephson junctions connected in parallel, allowing the supercurrents to interfere. The interference pattern is highly responsive to the magnetic flux passing through the loop, which is quantized due to the superconducting nature of the loop [24]. **Figure 2** illustrates the basic structure of a SQUID, which includes the superconducting loop, two Josephson junctions, and the direction of the supercurrent flow. The supercurrent is influenced by the magnetic field, which creates a voltage across the Josephson junctions. SQUIDs are renowned for their exceptional sensitivity in detecting minute changes in magnetic flux, making them indispensable in various fields, such as biomagnetism for medical diagnostics, geophysics for geological exploration, and fundamental physics for studying quantum phenomena and dark matter [25]. In essence, SQUIDs utilize the quantum properties of the Josephson Effect and flux quantization in superconductors to create highly sensitive magnetic field detectors, offering unparalleled detection capabilities for a wide range of applications [26]. Supercurrents in the loop are exquisitely sensitive to the presence of magnetic fields [27–29].

2.1 Superconducting transistors

Superconductors have the potential to revolutionize high-speed electronics and signal processing due to their unique properties. By eliminating electrical resistance, superconductors allow signals to propagate at incredible speeds without losing energy to heat, enabling faster data transmission and processing. This reduced resistance also leads to lower power consumption in superconducting circuits compared

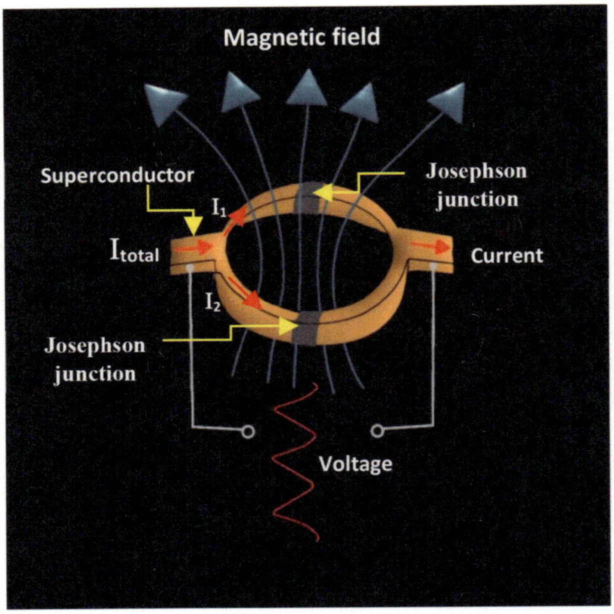

Figure 2.
Schematic illustration of a SQUID structure.

to conventional electronic circuits, resulting in improved efficiency [1, 30–32]. Superconducting transistors based on Josephson junctions can operate at extremely high frequencies, surpassing the capabilities of conventional transistors. This opens up possibilities for developing ultra-fast switching circuits, high-speed signal processing, and high-frequency communications. The Meissner effect in superconductors, which expels magnetic fields, reduces electromagnetic interference, leading to improved signal integrity, lower noise, and distortion in electronic circuits [33]. Superconducting interconnects can create high-speed data busses for faster communication between electronic components, while superconducting memories offer higher storage density and faster access speeds. Superconducting filters and mixers find applications in communication systems, radio astronomy, and advanced signal processing [34, 35]. Ongoing research aims to develop high-temperature superconductors that operate closer to room temperature, eliminating the need for extensive cooling systems. Integrating superconducting components with existing silicon-based electronics poses a significant challenge, with future efforts focusing on developing hybrid technologies that combine the benefits of both technologies [36].

2.2 Josephson junctions and their applications

Josephson junctions are at the core of many superconducting electronic devices, serving a pivotal role in leveraging the quantum properties of superconductors for diverse applications [37]. These junctions consist of two superconducting materials separated by a thin insulating barrier, typically a few nanometers thick, often made of an oxide or a normal metal. The distinguishing feature of Josephson junctions is the ability of Cooper pairs (two electrons bound together in a superconducting state) to tunnel quantum mechanically through the insulating barrier, defying classical barriers [38]. **Figure 3** provides a schematic representation of a Josephson junction,

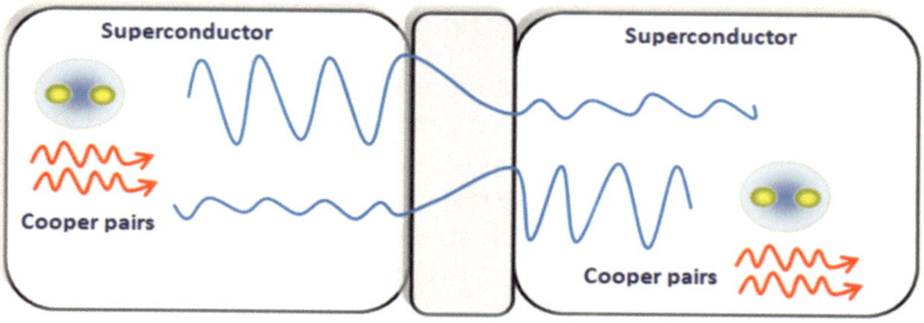

Figure 3.
Schematic illustration of a Josephson junction's structure.

illustrating the two superconducting materials separated by a thin insulating barrier. Within each superconductor, electrons pair up to form Cooper pairs, which are responsible for the superconducting properties. These Cooper pairs can tunnel through the insulating barrier, creating a supercurrent across the junction.

There are two key Josephson Effects: the DC Josephson Effect, where a supercurrent flows across the junction without any applied electric field when a voltage difference is present, and the AC Josephson Effect, where a constant voltage leads to supercurrent oscillations at a frequency proportional to the voltage. These effects are crucial for developing high-frequency devices [39].

Josephson junctions find applications in various superconducting electronics. For instance, SQUIDs (Superconducting Quantum Interference Devices) utilize Josephson junctions in a superconducting loop to create highly sensitive magnetometers used in MRI, biomagnetism, geophysics, and fundamental physics research. In quantum computing, Josephson junctions are essential for creating superconducting qubits, allowing for the manipulation and entanglement of quantum states. Moreover, these junctions are used in superconducting transistors for high-speed signal processing, digital circuits, and high-frequency communications [40]. The advantages of Josephson junctions lie in their unparalleled sensitivity, high-speed operation, and unique quantum effects. They offer high sensitivity for magnetic field detection in SQUIDs, enable ultra-fast switching in electronics, and exhibit quantum phenomena crucial for quantum computing applications [41]. However, challenges exist in the fabrication of these junctions with precise dimensions and properties, as well as the requirement for cryogenic cooling systems to operate at low temperatures.

3. Energy transmission and storage

Superconductors are transforming the way energy is transferred, particularly in power transmission lines. Traditional power lines made of copper or aluminum suffer from energy losses due to resistance, resulting in wasted energy and reduced efficiency. Superconductors, on the other hand, with their zero resistance at low temperatures, offer a solution to this problem.

3.1 Superconductors in power transmission lines

Superconductors offer significant advantages in power transmission lines by eliminating resistance, enabling energy transmission without losses due to heating [42].

This feature leads to substantial energy savings and a more efficient grid. Superconducting cables can carry higher current densities than traditional lines, allowing for the efficient transmission of more power over the same infrastructure, facilitating the transport of renewable energy sources over long distances [43, 44]. Additionally, the use of superconductors enables the deployment of smaller and less bulky cables, reducing the need for large towers and extensive right-of-way acquisition during installation [31]. This not only streamlines the installation process but also makes it more cost-effective, contributing to the overall efficiency and sustainability of the power transmission system. Superconducting transmission lines offer a promising solution to reduce energy losses and greenhouse gas emissions in power generation [30, 45]. Challenges hindering widespread implementation include the high costs of cooling systems, materials, and grid integration [46]. Ongoing research focuses on advancing high-temperature superconductors and more efficient cooling technologies to overcome these obstacles, paving the way for broader adoption of superconducting technology in power transmission [47]. The future of power grids envisions a cleaner, more efficient, and reliable energy supply, benefitting from the unique properties of superconductors, which can revolutionize the environmental impact and efficiency of power systems [44].

3.2 Superconducting magnetic energy storage (SMES) systems

Superconducting magnetic energy storage (SMES) systems are cutting-edge solutions for efficient energy storage, utilizing superconductors to store energy in a magnetic field within a superconducting coil [48]. These systems boast high-energy conversion efficiencies exceeding 90%, rapid energy storage and release in milliseconds, durable coils, and eco-friendliness with zero emissions during operation [49]. SMES applications span grid stabilization, renewable energy integration, power quality enhancement, backup power, and various power electronic uses like pulse generators and superconducting motors [50]. Additionally, SMES systems have been proposed as a reliable energy source for transportation systems, such as railways, by combining them with lithium batteries to ensure a stable and continuous power supply [51]. Despite challenges like high costs, limited storage capacity, and energy-intensive cooling systems, ongoing research focuses on cost-effective materials, improved coil designs, and reduced cooling energy needs [52]. Future advancements are expected to enhance SMES roles in grid stability, renewable energy integration, and overall energy sector reliability and efficiency [53].

4. Superconducting technology in medical imaging (MRI)

Superconducting technology has brought about a significant transformation in medical imaging, particularly in the realm of Magnetic Resonance Imaging (MRI). Superconducting magnets play a critical role in MRI machines by generating powerful magnetic fields that surpass the capabilities of conventional electromagnets [31, 54]. These strong magnetic fields are essential for producing high-resolution images of the human body, allowing for detailed visualization of internal structures and abnormalities. The stability and consistency of the magnetic field provided by superconducting magnets result in clearer and more detailed images, facilitating accurate diagnoses and treatment planning. This enhanced image quality is particularly beneficial for identifying subtle abnormalities, visualizing complex anatomical structures, and

Benefit	Description	Impact
Powerful magnetic fields	Superconducting magnets generate stronger magnetic fields compared to conventional electromagnets.	Enables high-resolution images of internal structures and abnormalities.
Improved image quality	Stable and consistent magnetic fields lead to clearer and more detailed images.	Facilitates accurate diagnoses, better treatment planning, and identification of subtle abnormalities.
Faster scan times	High magnetic field strength allows for quicker scans.	Reduces patient discomfort, anxiety, and increases patient throughput.
Specific applications	Neuroimaging, cardiovascular imaging, oncology.	Allows for detailed visualization of the brain, heart, and blood vessels, aids in cancer detection and monitoring.
Challenges	Cryogenic cooling systems, high manufacturing, and maintenance costs.	Limits accessibility and affordability of MRI technology.
Future directions	Research into high-temperature superconductors and cost reduction.	Aims to make MRI technology more accessible and affordable.

Table 1.
Superconducting technology in MRI: benefits and challenges.

monitoring disease progression [55]. Moreover, the high magnetic field strength of superconducting magnets enables faster scan times, reducing patient discomfort and anxiety during MRI procedures. Quicker scans also enhance patient throughput, enabling more individuals to benefit from this advanced imaging technology [56]. Superconducting MRI has specific applications in various medical fields, including neuroimaging for detailed brain visualization, cardiovascular imaging for assessing heart health, and oncology for cancer detection and treatment monitoring [48]. While superconducting technology has significantly advanced MRI, challenges remain in areas like the cost and maintenance of cooling systems and the overall manufacturing process. However, ongoing research efforts are focused on developing high-temperature superconductors and reducing costs to make MRI technology more accessible and affordable [50]. **Table 1** gives a summary of benefits, challenges, and future directions of superconducting MRI.

5. Conclusion

Superconductivity stands as a transformative technology with the potential to revolutionize diverse sectors, from computing to energy and medicine. Its unique properties, such as zero resistance and the Meissner effect, are being harnessed to create powerful, efficient, and innovative devices. Superconducting quantum computers, employing superconducting qubits and circuits, are poised to usher in an era of unprecedented computing power and speed. By addressing challenges like scaling and cost, these devices have the potential to solve complex problems in medicine, materials science, and artificial intelligence. In the energy sector, superconducting cables offer a path toward a more efficient and sustainable energy grid by enabling lossless energy transmission over long distances. Furthermore, superconducting magnetic energy storage (SMES) systems hold promise for grid stabilization and integration of renewable energy sources. The impact of superconductivity on medical imaging, particularly MRI, is already significant. Superconducting magnets enable high-resolution imaging and faster scan times, leading to more accurate diagnoses and improved patient care. Despite challenges such as cost, material limitations, and the need for cryogenic cooling, ongoing research endeavors are continually pushing the boundaries of superconducting technology. The development of high-temperature superconductors, more efficient cooling systems, and improved fabrication techniques are promising avenues for overcoming these hurdles.

Acknowledgements

The author acknowledges the use of Google's Gemini for language polishing and translation of the manuscript.

Author details

Belqees Hassan[1,2]

1 Department of Physics, College of Science, Qassim University, Buraydah, Saudi Arabia

2 Department of Physics, Faculty of Science, Ibb University, Ibb, Yemen

*Address all correspondence to: b.hassan@qu.edu.sa;mailto; belqees.ahmad@yahoo.com

References

[1] Rizwan M, Ayub A, Fatima S, Ilyas I, Usman A, Shoukat A. Fundamentals and properties of superconductors. Materials Research Foundations. 2022;**132**:49-78. DOI: 10.21741/9781644902110-3

[2] Wang C. Theory and application of superconducting materials. In: Theory and Application of Rare Earth Materials. Singapore: Springer; 2023. DOI: 10.1007/978-981-19-4178-8_14

[3] Wu H. Recent development in high temperature superconductor: Principle, materials, and applications. Applied and Computational Engineering. 2024;**63**:153-171. DOI: 10.54254/2755-2721/63/20241015

[4] Zonno M, Yi M. Probing unconventional superconductivity using synchrotron radiation. Synchrotron Radiation News. 2023;**36**(3):2-3. DOI: 10.1080/08940886.2023.2226050

[5] Latif S, Muhammad Husnain M, Siddique H, Imran M. Basic concepts and properties of superconductors. Materials Research Foundations. 2022;**132**:1-16. DOI: 10.21741/9781644902110-1

[6] Hasan MS, Ali SS. Properties and types of superconductors. Materials Research Foundations. 2022;**132**:17-48. DOI: 10.21741/9781644902110-2

[7] Krasnok A, Dhakal P, Fedorov A, Frigola P, Kelly M, Kutsaev S. Superconducting microwave cavities and qubits for quantum information systems. Applied Physics Reviews. 2024;**11**:011302. DOI: 10.1063/5.0155213

[8] Psaroudaki C, Peraticos E, Panagopoulos C. Skyrmion qubits: Challenges for future quantum computing applications. Applied Physics Letters. 2023;**123**:260501. DOI: 10.1063/5.0177864

[9] Dejpasand MT, Sasani Ghamsari M. Research trends in quantum computers by focusing on qubits as their building blocks. Quantum Reports. 2023;**5**:597-608. DOI: 10.3390/quantum5030039

[10] Wu A, Yin K, Cross AW, Li A, Ding Y. Enabling Full-Stack Quantum Computing with Changeable Error-Corrected Qubits. 2023. DOI: 10.48550/arXiv.2305.07072

[11] Wei X, Jiang J, Xu W, Guo T, Zhang K, Li Z, et al. Compact superconducting transmon qubit circuits made of ultrathin NbN. Applied Physics Letters. 2023;**123**:154005. DOI: 10.1063/5.0170259

[12] Kartal G, Simion G, Sorée B. A Superconducting Quantum Information Processor with High Qubit Connectivity. 2023. DOI: 10.48550/arXiv.2307.08101

[13] Shirai S, Okubo Y, Matsuura K, Osada A, Nakamura Y, Noguchi A. All-microwave manipulation of superconducting qubits with a fixed-frequency transmon coupler. Physical Review Letters. 2023;**130**:260601. DOI: 10.1103/PhysRevLett.130.260601

[14] Hassani F, Peruzzo M, Kapoor LN, et al. Inductively shunted transmons exhibit noise insensitive plasmon states and a fluxon decay exceeding 3 hours. Nature Communications. 2023;**14**:3968. DOI: 10.1038/s41467-023-39656-2

[15] Spethmann M, Bosco S, Hofmann A, Klinovaja J, Loss D. High-fidelity two-qubit gates of hybrid superconducting-semiconducting singlet-triplet.

Physical Review B. 2024;**109**:085303. DOI: 10.1103/PhysRevB.109.085303

[16] Jin Z, Li S, Wang X, Liang F, Peng CZ. A co-simulation of superconducting qubit and control electronics for quantum computing. Review of Scientific Instruments. 2023;**94**(10). DOI: 10.1063/5.0163725

[17] Mishra H, Bonam S, Pandey U, Singh SG. Thermally annealed tantalum-filled vertical superconducting interconnects for scalable quantum computing systems. In: 2023 IEEE 25th Electronics Packaging Technology Conference (EPTC). Singapore; 2023. pp. 498-503. DOI: 10.1109/EPTC59621.2023.10457669

[18] Last T, Mongillo M, Ivanov T, Rol A, Lawrence A, Alberts G, et al. Key ingredients for manufacturing superconducting quantum processors at scale. In: Proc. SPIE 12497, Novel Patterning Technologies. 2023. p. 124970H. DOI: 10.1117/12.2657319

[19] Kapit E, Oganesyan V. Small Logical Qubit Architecture Based on Strong Interactions and Many-Body Dynamical Decoupling. 2022. DOI: 10.48550/arXiv.2212.04588

[20] Yang P et al. Superconducting quantum interference device magnetometer system for geomagnetic field and VLF field measurements. In: 2022 IEEE MTT-S International Microwave Workshop Series on Advanced Materials and Processes for RF and THz Applications (IMWS-AMP), Guangzhou, China. 2022. pp. 1-3. DOI: 10.1109/IMWS-AMP54652.2022.10107112

[21] Vettoliere A, Silvestrini P, Granata C. Superconducting quantum magnetic sensing. Quantum Materials, Devices, and Applications. 2023:43-85. DOI: 10.1016/b978-0-12-820566-2.00001-6

[22] Hajalilou A, Tavakoli M, Parvini E. Insight into superconducting quantum interference devices (SQUID). In: Hajalilou A, Tavakoli M, Parvini E, editors. Magnetic Nanoparticles. 2022. DOI: 10.1002/9783527840762.ch6

[23] Bi X, Tian F, Chen G, Li Z, Qin F, Lv YY, et al. A superconducting micro-magnetometer for quantum vortex in superconducting nanoflakes. Advanced Materials. 2023;**35**(19):e2211409. DOI: 10.1002/adma.202211409

[24] Wang Y. Unified Network Equation for Both Normal RLC Circuits and Josephson Junction Circuits. arXiv:2109.12476 [cond-mat.supr-con]. 2021. DOI: 10.48550/arXiv.2109.12476

[25] Noh T, Kindseth A, Chandrasekhar V. Nonlocal superconducting quantum interference device. Physical Review B. 2021;**104**(6):064503. DOI: 10.1103/PhysRevB.104.064503

[26] Karimov T, Ostrovskii V, Rybin V, Druzhina O, Kolev G, Butusov D. Magnetic flux sensor based on spiking neurons with josephson junctions. Sensors. 2024;**24**(7):2367. DOI: 10.3390/s24072367

[27] Flokstra J, Seidel P. SQUIDs. In: Handbook of Superconductivity. CRC Press; 2022. pp. 672-681. DOI: 10.1201/9781003139638

[28] Schmelz M, Stolz R. Superconducting quantum interference device (SQUID) magnetometers. In: Grosz A, Haji-Sheikh M, Mukhopadhyay S, editors. High Sensitivity Magnetometers. Smart Sensors, Measurement and Instrumentation. Vol. 19. Cham: Springer; 2017. DOI: 10.1007/978-3-319-34070-8_10

[29] Drung D, Beyer J. Application in superconducting quantum interference devices SQUIDs. In: Josephson Junctions. Jenny Stanford Publishing; 2017. pp. 245-329. DOI: 10.1201/9781315364520-8

[30] Molodyk A, Larbalestier DC. The prospects of high-temperature superconductors. Science. 2023;**380**:1220-1222. DOI: 10.1126/science.abq4137

[31] Ali SS, Zulqarnain M. Superconductors for medical applications. Materials Research Foundations. 2022;**132**:211-229. DOI: 10.21741/9781644902110-12

[32] Bose S, Chakraborty S, Roy B, Nath PC. Superconductors for large-scale applications. Materials Research Foundations. 2022;**132**:79-96. DOI: 10.21741/9781644902110-4

[33] Basu S, Chattopadhyay S, Superconductivity, Modern Perspectives in the Study of Electronic Systems. 2022. DOI: 10.1063/9780735422537_004

[34] Tanaka M, Yamanashi Y. Current status and issues of superconductor peripheral circuits for quantum computers. Denki Gakkai Ronbunshi. A (Online). 2022;**142**(5):175-182. DOI: 10.1541/ieejfms.142.175

[35] Riva N, Grilli F, Dutoit B. Superconductors for power applications: an executable and web application to learn about resistive fault current limiters. European Journal of Physics. 2021;**42**(4):045802. DOI: 10.1088/1361-6404/ABF0DA

[36] Elmgren K, Acharya A, Hunt W. Superconductor Electronics Research: National Competitiveness and Funding Activity. Center for Security and Emerging Technology; 2021. DOI: 10.51593/20210003

[37] Gallop JC. SQUIDs, the Josephson Effects and Superconducting Electronics. 1st ed. CRC Press; 1991. DOI: 10.1201/9780203738887

[38] Plastovets V, Mel'nikov AS. Electronic structure of a Josephson vortex in a SIS junction. Physical Review B. 2022;**105**:094516. DOI: 10.1103/PhysRevB.105.094516

[39] Gallop J, Braginski AI. Introduction to section H3: Josephson junction devices. In: Handbook of Superconductivity. CRC Press; 2022. pp. 640-641. DOI: 10.1201/9781003139638

[40] Wang Y. Magnetic-flux-flow diagrams for design and analysis of josephson junction circuits. In: IEEE Transactions on Applied Superconductivity. Vol. 33. Issue. 7. 2023. pp. 1-8. Art no. 1400508. DOI: 10.1109/TASC.2023.3290199

[41] Walsh ED, Jung W, Lee GH, Efetov DK, Wu BI, Huang KF, et al. Josephson junction infrared single-photon detector. Science. 2021;**372**(6540):409-412. DOI: 10.1126/science.abf5539

[42] Reiser W, Reek T, Hanebeck C, Abrell P. Superconductor busbar systems in the light of increased energy costs. In: Alam S et al., editors. Energy Technology 2023. TMS 2023. The Minerals, Metals and Materials Series. Cham: Springer; 2023. DOI: 10.1007/978-3-031-22638-0_18

[43] Navneet K, Mona KR, Singh J, Rana S. Superconductors for energy storage. Electroceramics for High Performance Supercapacitors. 2023;**2023**:95-134. DOI: 10.1002/9781394167166.ch5

[44] Miúdo L, Pina JM, Amaro N. Superconducting saturable core reactor as variable inductance for controlling

the power flow in a transmission line. In: Camarinha-Matos LM, Ferrada F, editors. Technological Innovation for Connected Cyber Physical Spaces. DoCEIS 2023. IFIP Advances in Information and Communication Technology. Vol. 678. Cham: Springer; 2023. DOI: 10.1007/978-3-031-36007-7_6

[45] Vijayapakavan P, Smart DR, Ramu K, Ramachandran M. Superconducting electromagnetic launch machine system for aerospace applications. Journal on Applied and Chemical Physics. 2023;**2**:40-47. DOI: 10.46632/jacp/2/1/5

[46] Paramane A, Chandrasekaran T, Nazir MT. A review on insulation and dielectrics for high temperature superconducting cables for power distribution: progress, challenges, and prospects. IEEE Transactions on Applied Superconductivity. 2023;**33**(6):1-31. DOI: 10.1109/TASC.2023.3267055

[47] Jacob T, Buchholz A, Noe M, Weil M. Comparative life cycle assessment of different cooling systems for high-temperature superconducting power cables. In: IEEE Transactions on Applied Superconductivity. Vol. 32. Issue 4. 2022. pp. 1-5. Art no. 4802805. DOI: 10.1109/TASC.2022.3168239

[48] Ghiasi S, Mohammad Sadegh Ghiasi S. A Review on Superconducting Magnetic Energy Storage System Applications. London, UK: IntechOpen; 2023. DOI: 10.5772/intechopen.110889

[49] Ali MH. Superconducting Magnetic Energy Storage in Power Grids. USA: Electrical and Computer Engineering Department, Electric Power and Energy Systems Laboratory, University of Memphis; 2022. DOI: 10.1049/pbpo210e

[50] Lu R. Sustainability and environmental efficiency of superconducting magnetic energy storage (SMES) technology. Highlights in Science, Engineering and Technology. 2022;**26**:365-371. DOI: 10.54097/hset.v26i.4005

[51] Ikram M, Raza A, Altaf S, Ahmed Rafi A, Naz M, Ali S, et al. High Temperature Superconductors. London, UK: IntechOpen; 2021. DOI: 10.5772/intechopen.96419

[52] Shen B, Chen Y, Fu L, Xu J, Chen X, Zhang M. Superconducting Magnetic Energy Storage (SMES) for Railway System. In: 2023 IEEE International Conference on Applied Superconductivity and Electromagnetic Devices (ASEMD). Vol. 2023. Tianjin, China. 2023. pp. 1-2. DOI: 10.1109/ASEMD59061.2023.10369041

[53] Cazacu D. 'Overview of SMES technology' energy engineering. In: Superconducting Magnetic Energy Storage in Power Grids. Chapter 2. IET Digital Library; 2022. pp. 11-57. DOI: 10.1049/PBPO210E_ch2. Available from: https://digital-library.theiet.org/content/books/10.1049/pbpo210e_ch2

[54] Yokoyama S. Technical overview of superconducting magnets. Japanese Society for Magnetic Resonance in Medicine. 2022. DOI: 10.2463/jjmrm.2022-1771

[55] Raza A, Ali SS. Superconductors for magnetic imaging resonance applications. Materials Research Foundations. 2022;**132**:230-255. DOI: 10.21741/9781644902110-13

[56] Parizh M, Stautner W. MRI magnets. In: Handbook of Superconductivity. CRC Press; 2022. pp. 437-492. DOI: 10.1201/9781003139638

Chapter 6

Superconductor Magnets

Habtamu Anagaw and Gedefaw Mebratie

Abstract

Superconductors are materials that conduct electricity without any resistance, offering a world of potential applications. At extremely low temperatures, their electrical resistance becomes zero. This property enables them to carry massive currents, making them ideal for technologies like maglev trains, improved power grids, medical devices, and particle accelerators. While superconducting technology has advanced significantly, MRI and NMR are currently the main commercial applications. High-temperature superconductors (HTS) with transition temperatures in excess of 100 K have renewed interest in using superconductivity in power technology. While low-temperature superconductors (LTS) were previously explored, they did not lead to commercial products. This book reviews past and present activities in power generation, transmission, and distribution using HTS. Many groups are working on superconducting generators, motors, and cables, which could lead to significant growth in superconductor applications. However, technological and economic challenges with HTS materials must be overcome. The review chapter covers the properties of HTS, cooling systems, and applications in devices like rotating machines, transformers, and maglev trains. It also provides examples of device designs for practical use.

Keywords: superconductor, quench, cryogenics, homogeneity, Dewar, magnetic energy storage, MRI, particle accelerator, maglev

1. Introduction

1.1 Discovery and development of superconductivity

Superconductivity is the spectacular phenomenon in which a material makes a second-order phase transition from a conventional metal to an exotic substance at a specific temperature, known as the superconducting transition temperature T_C, allowing electrical current to flow without resistance. Heike Kamerlingh Onnes and his graduate student, G. Holst, discovered this intriguing new phenomenon in 1911 in Leiden, Netherlands by cooling mercury to liquid helium temperature using liquid helium as a refrigerator. The temperature-dependent resistivity was observed to drastically drop at 4.2 K, indicating the commencement of a phase transition to the superconducting state, when the electrical resistance abruptly became zero [1]. Superconductivity is now a well-known quantum phase of matter and has been discovered to exist in different materials, ranging from elements and alloys to complex compounds and even

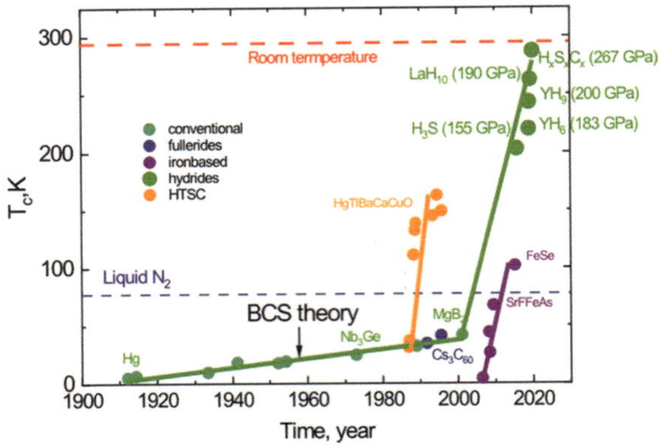

Figure 1.
Timeline of the discovery of superconductors [5].

organic molecules. Superconductivity can be accompanied by other intriguing and different phases in a number of these systems, such as magnetism, charge ordering, and quantum criticality, adding to the variety of these materials.

After the Meissner effect was discovered, the London brothers Fritz and Heinz developed a phenomenological justification motivated by concepts from the two-fluid model of superfluid ^4He [2]. Despite its simplicity, the theory was able to accommodate the Meissner effect as well as make a number of other correct predictions about the electrodynamics of a superconductor [3]. Superconductivity was thought to be more or less known for a number of years following the BCS theory [4].

The history of superconductors goes back to 1911, when Kamerlingh Onnes discovered the phenomenon of superconductivity in mercury. Subsequently, a number of superconducting materials were discovered, as shown in **Figure 1**.

1.2 The BCS theory

In 1957, Bardeen, Cooper, and Schrieffer (BCS) postulated a microscopic theory of superconductivity based on the fundamental quantum mechanics principles. According to the BCS theory, superconductivity is based on the presence of attractive forces between electrons caused by electron-phonon interactions rather than repulsive forces between electrons. The presence of an attracting force results in superconducting characteristics. When two electrons have propagation vectors that are identical in magnitude and opposite in direction, the attractive force becomes maximal. The core claim of the BCS theory is that two electrons in a spin-singlet state with antiparallel spins can create a bound state, known as a Cooper pair. The primary findings of the BCS theory described the superconducting energy gap inside an s-wave paradigm (which is typical for many low-temperature superconductors but not established in unconventional superconductors such as d-wave high-temperature superconductors). Furthermore, the BCS theory can approximate the quantum-mechanical many-body state of the system of electrons within the metal [6].

1.3 Heavy fermion materials

Materials containing actinide and lanthanide atoms, with partially filled f shells, are classified as heavy fermion metals. It is possible to speculate that the local moments of the f-electrons in heavy fermion metals as a Fermi gas at room temperature because of their weak interaction with the conduction electrons. Heavy fermion metals, on the other hand, are generally defined as those at lower temperatures where there is high coupling between the f-electrons and conduction cloud (Kondo effect) [7].

The partially filled f-orbitals of rare-earth or actinide ions are the source of the features of heavy-fermion compounds [8]. The f-orbitals of these ions collapse within the ions of the inert gas core due to their enormous nuclear charge, converting them into localized magnetic moments. Similar peculiar superconducting phases are even present in some of these heavy-fermion materials. This class of heavy-fermion materials can be broadly defined as systems that are near a magnetic instability and possess the characteristic features of showing evidence of local moments at high temperatures and resembling a Fermi liquid with very heavy quasi-particle masses at low temperatures. The lanthanide or actinide series of elements, which have incomplete f shells, provide the basis for the heavy-fermion materials. Because of the tiny radius of the ionic f-orbitals, the f-derived electronic states are nearly localized and exhibit strong electronic interactions, which preserves much of their ionic character. The strong ionic coulomb interactions, which tend to localize electrons and produce local magnetic moments, and the hybridization with extended band states, which tend to delocalize the electrons, are delicately balanced in heavy-fermion materials. The temperature-dependent transition from the high-temperature local moment regime to the low-temperature zone of itinerant f electron behavior is caused by the delicate balance [9].

1.4 High-temperature superconductivity in copper oxides

The discovery of high-temperature superconductivity (HTS) in copper oxides in 1986 marked a new era of superconductivity. Unlike traditional superconductors, these HTS materials have a much higher upper critical field (H_{C2}), allowing for measurements in extremely high magnetic fields. In the “mixed” state between H_{C1} and H_{C2}, superconductivity and magnetic fields coexist. This state is dominant in HTS materials, while the Meissner state (where magnetic fields are completely expelled) is only present at very low fields. Importantly, the critical temperature (T_C) in HTS systems is comparable to the Fermi energy (E_F). The higher the Fermi energy, the higher the T_C value, which challenges the conventional BCS theory, a widely accepted theory of superconductivity.

Due to the high T_C of cuprates, scientists have been working to study their behavior in extremely high magnetic fields. This research is crucial for understanding superconducting fluctuations, vortex arrays, and quantum oscillations in these materials. Recent discoveries of HTS in iron pnictides and chalcogenides, which are less anisotropic than cuprates, have renewed interest in this field. Cuprate high-temperature superconductors (HTS) are different from conventional BCS superconductors. BCS superconductors pair electrons due to attractive interactions mediated by lattice vibrations, while cuprates likely pair electrons due to attractive interactions related to magnetic field energy. This viewpoint is supported by the unconventional d-wave nature of Cooper pairs in cuprates. BCS theory suggests that superconducting pairing affects only a small fraction of electrons near the Fermi surface. In conventional BCS superconductors, Cooper pairs form in an s-wave state, leading to a uniform energy gap around the Fermi surface. However, in cuprates, certain points or lines on the Fermi surface have no energy gap, leading to significantly different thermodynamic and transport properties compared to BCS theory.

To understand deviations from the mean-field BCS description of high-temperature superconductors (HTS), researchers are focusing on low-energy quasi-particles and their interactions with superconducting vortices, particularly through thermal and quantum fluctuations. High magnetic fields are essential for weakening superconductivity and inducing strong fluctuations. This research involves studying the phase diagram of HTS, and various elements of this problem are being investigated intensively [10].

High magnetic fields are crucial for superconductivity research, especially in cuprates and pnictides. They allow scientists to explore the low-temperature regime where quantum fluctuations are significant. This can lead to the discovery of new quantum states of matter that are not simply uncovered but actually induced by high magnetic fields. One example of such a state is the formation of unidirectional charge- or spin-ordered regions in cuprates, known as stripes [11]. Cuprate high-temperature superconductors are the only suitable material for creating all-superconducting magnets at 30 T and above.

1.5 High-temperature superconductivity in iron pnictides and chalcogenides

Iron-based HTS was discovered in 2008, marking a significant breakthrough as the first non-copper-based HTS materials. This discovery came after the 2005 nonreciprocal critical current (NRC) study on Opportunities in High Magnetic Field Science [12]. Iron-based HTS is a rapidly growing field of research. These materials have multiple bands and a Fermi surface with several distinct "pockets." The superconducting gap is believed to have opposite signs on these pockets, suggesting that interelectron repulsion is involved in the superconductivity mechanism. Both cuprate and pnictide families share several common characteristics [13]. Both cuprate and pnictide superconductors exhibit strong electronic correlations in their normal state [14]. However, pnictides are less anisotropic than cuprates. The pnictides show great promise for high-field applications, because their relatively high T_C, large critical fields, and weaker anisotropies are appealing virtues for such applications [15].

1.6 Organic superconductors

Organic superconductors were among the first materials studied in high magnetic field research. While not considered high-temperature superconductors, they are extreme type II systems with unique anisotropic properties. High magnetic fields allow researchers to explore regimes similar to those found in cuprates and pnictides. Phenomena such as the competition between superconductivity and spin/charge density waves, quantum oscillations in the superconducting state, and other interesting behaviors were first observed in organic superconductors. This area of research remains active, with high magnetic fields continuing to play a crucial role [16].

2. Superconducting magnet

A superconducting magnet is an electromagnet that uses a superconducting coil. Unlike traditional electromagnets, superconducting magnets can vary their field strength without generating heat, allowing for higher current densities and stronger magnetic fields. The development of technical superconductors in the 1960s made

DOI: http://dx.doi.org/10.5772/intechopen.1007576

superconducting magnets possible. Since then, superconducting magnet technology has become a major global industry and the most important application of superconductivity [17].

Superconducting magnets allow high-field solenoids with no steady-state power consumption, but they do need very low temperatures. Early superconducting materials had such a low critical temperature, T_C, that they had to be cooled using liquid helium (boiling point 4.2 K), an expensive undertaking. More recently, superconducting oxides have been developed with critical temperatures above the boiling point of liquid nitrogen (77.4 K), making superconducting magnets much more practical. The high-T_C superconductors also have a high critical field—the magnetic field strength that quenches superconduction that sets an upper limit to the performance of a superconductor [18].

Superconducting magnet technology deals with the design, manufacture, and operation of superconducting magnets. A superconducting magnet is a highly stressed device: it requires the best that engineering has to offer to ensure that it operates successfully, is reliable, and at the same time is economically viable. A typical 10-tesla magnet is subjected to an equivalent magnetic pressure of 40 MPa (nearly 400 atm), whether it operates superconductively at 4.2 K (liquid helium) or 77 K (liquid nitrogen), or resistively at room temperature. Superconducting magnet technology is interdisciplinary in that it requires knowledge and training in many fields of engineering—mechanical, electrical, cryogenic, and materials [19].

Superconducting magnets are essential for applications like particle accelerators and fusion reactors. Key concepts related to their operation include quenches (loss of superconductivity), cryogenics (cooling), field homogeneity (uniformity), and Dewar (a vacuum flask).

2.1 Quench

A quench occurs when a superconducting magnet unexpectedly loses its superconductivity, often due to heat or other factors. This can cause a significant release of thermal energy, making it important to have strategies in place to manage the situation and minimize costs and downtime [20]. A quench is a critical failure in superconducting structures that occurs when parts of the superconducting coil lose their superconductivity. This leads to voltage, temperature rise, mechanical stress, increased cryogen pressure, and the expulsion of magnetic flux. Early detection and protection of quenches are essential for the safe operation of superconducting structures, especially high-temperature superconductors (HTS) which have high critical temperatures and can carry large currents. Researchers have focused on improving the stability of superconducting structures and magnets to prevent damage caused by local quenches [21].

To design and develop effective quench detection and protection methods for superconducting structures, it is crucial to understand their electrical, thermal, and structural behavior during a quench. Researchers have reported various experimental methods for quench detection, including acoustic emission detection [22], the ultrasonic wave method [23], the resistive voltage detection method [24], the strain-based detection method for LTS solenoids [25], as well as the new quench detection technique based on Rayleigh-backscattering interrogated optical fibers for HTS magnets [26]. A simple technique for quench detection in superconducting coils and magnets is individual voltage monitoring. Furthermore, a number of theoretical frameworks and numerical simulations have been put forth to examine quenches and advance knowledge on quench protection and detection in both high- and low-temperature superconducting systems [27]. Using equivalent network circuits and homogenized coil models, researchers have developed numerical methods and computer programs to simulate quenches in LTS coils and magnets [28].

2.2 Cryogenics

At cryogenic temperatures of 4–80 K, all superconductors that are now known to exist must function. Building cooling devices—also referred to as refrigerators or simply cryocoolers—that can achieve extremely low temperatures requires a significant deal of inventiveness since the technology involved in obtaining such low temperatures is quite complex. Superconductor magnet coils are kept and stored in a cryostat in a temperature and humidity-controlled environment an assembly of cryostats to hold the HTS magnet. To keep this HTS magnet working at the required temperature of 30 K, it is conduction-cooled using a cryocooler and epoxy-impregnated.

The vacuum vessel, HTS magnet, current leads, cold mass support, intermediate temperature radiation shield, multi-layer insulation (MLI), and cryocooler make up the cryostat assembly. The HTS magnet is placed inside the vacuum vessel, which also has a transparent 100 mm diameter bore. The heated vacuum vessel provides support for the HTS magnet coil assembly, also referred to as the "cold mass." The magnet assembly is surrounded by a shield of intermediate temperature that is kept at 60 K. This shield's job is to block heat radiation coming from the vacuum vessel's room-temperature wall. In the vacuum vessel's area between the radiation shield and room-temperature walls, a multi-layer insulation (MLI) thermal blanket is attached. The magnet receives current from two leads. Since one of the main sources of heat conduction is current leads [29].

Common current leads consist of two sections: an HTS-made lower section that is positioned between the magnet and radiation shield, and an upper copper or brass section that is positioned between the radiation shield and ambient temperature. The first stage of the cryocooler intercepts the thermal load transmitted from room temperature to the radiation shield. Analogously, the second cryocooler stage eliminates the thermal load transferred from the radiation shield to the magnet. For the purpose of cooling the magnet assembly, a two-stage cryocooler is shown in **Figure 2**. Through the MLI thermal insulation, conduction through the cold mass supports, instrumentation wires, and current leads at the chosen 60 K radiation shield, the first stage cooling intercepts the thermal load radiated from room temperature. Thermal radiation from the 60 K shield, thermal conduction through cold mass supports, instrumentation wires, and conduction down HTS current leads between the magnet cold mass and the cryocooler's first stage are all eliminated by second-stage cooling, which eliminates thermal burden in the 35 K region. It has even been possible to chill a magnet to 4 K using two-stage cryocoolers, which is required for LTS magnets [29].

2.2.1 Cryogenic fluids for cooling HTS magnets

Cryogenics uses cryogenic fluids like liquid helium to keep superconducting magnets at low temperatures. During a quench, the cryogenic system must effectively remove the coolant to prevent excessive pressure buildup, ensuring the safe operation of the superconducting magnets. Cryogenics is vital for the operation and efficiency of superconducting magnets, used in applications like particle accelerators and magnetic levitation. Cryogenics in these systems is characterized by cooling methods, thermal management, and operational stability [30].

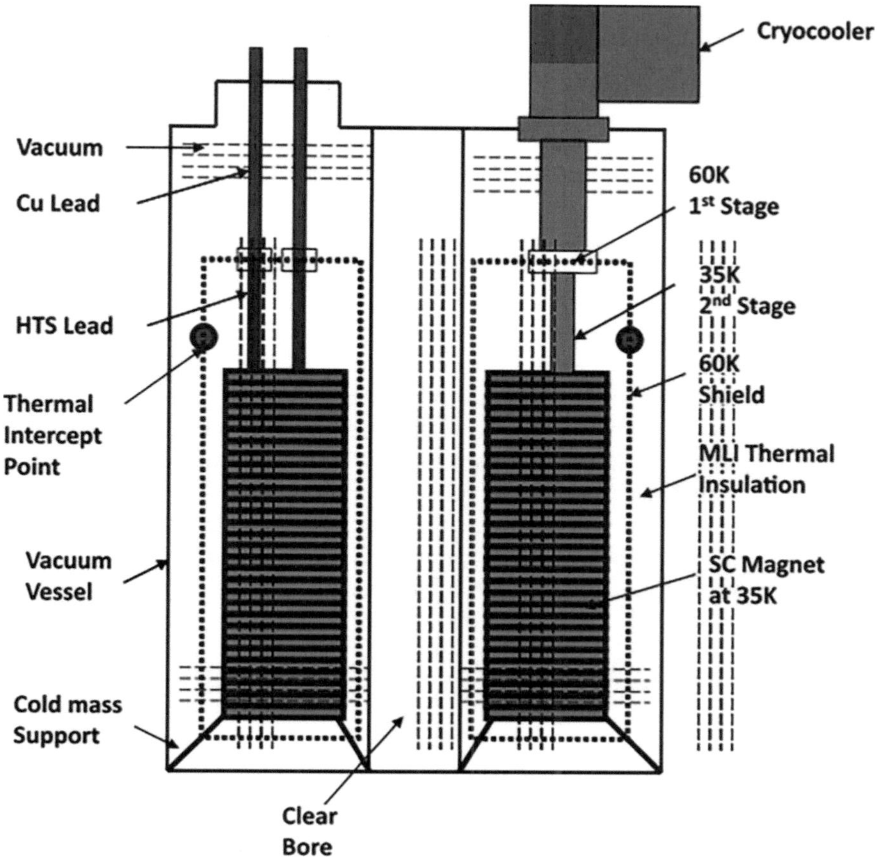

Figure 2.
Schematic of a cryostat.

It is necessary to cool a superconducting magnet to a temperature that allows for dependable and effective functioning. Bathing a magnet in a pool of liquid cryogen, which boils at a temperature appropriate for superconductor materials, is the most straightforward method of cooling it. If a cryogen is kept at constant pressure, it will boil at a constant temperature. Liquid He (LHe) has been the only fluid of choice because the majority of LTS magnets constructed using niobium-titanium (Nb-Ti) and niobium-tin (Nb_3Sn) superconductors need to operate at temperatures close to 4 K. The coolant used in nearly all high-field LTS magnets is LHe. Nonetheless, for low-field applications, HTS conductors (BSCCO-2223 and YBCO-123) can function at LN_2 temperature (77 K), and for high-field applications, at LNe temperature (27 K). Because it is cheap, inert, and easily accessible, LN_2 is favored as a coolant. While pricey and not easily accessible in significant amounts, LNe is also a desirable coolant for HTS magnets. All cryogens have the drawback of needing to be refilled or utilized in a closed cycle where a cryocooler recondenses boiled-off vapor. LNe and LN_2 have been employed as coolants in numerous HTS magnet applications; they will be covered in the upcoming chapters. Because of the risks involved, it is imperative that all safety and control procedures are followed when working with cryogens [31].

2.2.2 Direct cooling with cryogens

Superconductor LTS magnets using NbTi and Nb_3Sn are typically cooled by immersion in LHe. This is how almost all NbTi-built magnetic resonance imaging (MRI) magnets are cooled. Any heat produced in the NbTi conductor is passed to the liquid coolant since it is submerged in LHe. By applying the latent heat of vaporization, LHe is transformed into gaseous He (GHe), absorbing this heat load. As long as the heat flux—or heat per unit area—available at the conductor surface is less than what the LHe can remove, superconductors in these magnets will remain in the superconducting state. This limit is known as the critical heat flux. The superconductor temperature rises when the critical heat flow is exceeded, and eventually, when the temperature rises above the critical temperature (T_C), the superconductor transforms to its normal state. This group includes most MRI and particle accelerator magnets [29].

A forced flow of liquid helium cools another class of magnets. Using this technique, superconductor strands are either connected to the conduit's exterior or kept inside. In order to cool the superconductor strands, LHe is pushed through the conduit. Compared to the pool boiling process, this approach allows for a more regulated and efficient cooling of superconductor strands. Both methods have been used to create very massive magnets. The reader is referred to a book by Wilson for a comprehensive analysis of the behavior of LTS superconductors employed as magnets [32].

2.2.3 Indirect or conduction cooling

Cryocooler freezers have made conduction cooling easy and convenient for many high-field magnet construction projects. Commercially available magnets beyond 12 T are currently accessible. Epoxy is used to coat superconducting coils, resulting in monolithic constructions that are both easily handled and structurally robust. Conduction cools these coils from their exterior surface. **Figure 3** depicts two common conduction cooling techniques in which good thermal contact is achieved by (1) tubes transporting coolant and (2) a cryocooler refrigerator cooler directly [29].

A liquid cryogen or a cold gas could be the coolant in the cooling tubes. In a refrigerator or other comparable cooler, the coolant can function in a closed cycle by rejecting its heat load outside the magnet system. A cryogen extracted from a storage container could alternatively be used as the coolant. Furthermore, an HTS coil that is cooling could be thermally interfaced to a cryocooler refrigerator [29].

2.2.4 Refrigeration systems

The process of rejecting heat from a substance or an enclosed area and transferring it to another location with the primary goal of reducing and subsequently preserving

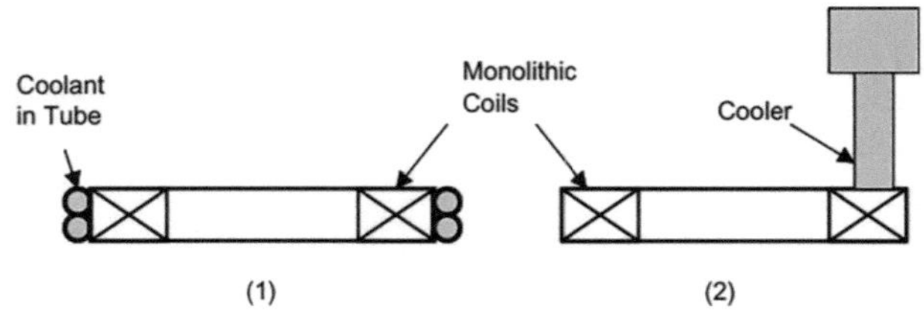

Figure 3.
Conduction cooling of a superconducting coil: (1) Coolant in tubes and (2) coil in contact with a refrigerator cooler.

the temperature of the substance or confined space is known as refrigeration. Power is needed by refrigerator systems in order to remove the thermal burden from the low-temperature area [29].

Cooling mechanisms: Superconducting magnets need to be kept very cold to work properly. The Super KEKB accelerator uses a cryogenic system that cools its magnets to 4.41 K using a 250-W refrigerator. This helps to minimize heat leaks and keep the system running smoothly [33]. HTS magnets use advanced cooling systems like Stirling-type pulse tube cryocoolers. These cryocoolers can achieve cooling capacities of 100 W at 45 K, making HTS magnets more energy-efficient [34].

Thermal management: Effective insulation and heat management are crucial to prevent heat from entering superconducting systems. The design of cryostats (containers for storing cryogenic fluids) is important for maintaining low temperatures and efficient operation. Compact testing facilities have been developed to measure heat loads and temperature changes in superconducting magnets, enabling precise control and analysis during operation [35].

2.3 Homogeneity

In superconducting magnets, homogeneity refers to how uniform the magnetic field is. Achieving high homogeneity is important for applications like MRI and particle physics, as field variations can affect their performance. It is essential in superconducting magnets for applications like MRI and particle accelerators, as it ensures uniform magnetic fields and improves performance. Recent studies have identified factors that affect homogeneity, such as thermal strains, structural design, and shielding techniques [36].

High homogeneity is crucial for accurate measurements in medical imaging and particle physics. In superconducting magnets, especially those operating at very low temperatures, maintaining field uniformity is important despite the thermal strains that occur during cooling [37].

High homogeneity magnets often use multiple solenoids and advanced materials like HTS to improve their performance [38]. In order to optimize the structural integrity and performance of these magnets under operational pressures, numerical homogenization approaches are utilized to simulate and anticipate their electromagnetic characteristics [39].

2.3.1 Challenges and solutions

The degradation of magnetic field homogeneity during cool-down might be caused by thermal stresses; therefore, designs must take these effects into consideration [37]. Novel techniques that eliminate undesired magnetic field components, including the use of superconducting shields, can significantly increase field homogeneity. While achieving high homogeneity is critical, it is also essential to balance this with the operational stability and safety of superconducting magnets, particularly under varying thermal and mechanical conditions [40].

2.4 Dewar

Cryogenic fluids are stored in Dewars, which are specialized containers with insulation to preserve low temperatures. It is essential to maintain the conditions required for superconductivity in the cryogenic system of superconducting. For superconducting magnets to function, the Dewar is an essential part that keeps the temperature down, which makes the magnets superconductive. Superconducting systems' performance and dependability are greatly impacted by their characteristics and design magnets [41].

2.4.1 Functionality and design

2.4.1.1 Thermal insulation

Dewars are made to keep heat transfer to a minimum. To maintain the low temperatures required for superconducting magnets, a vacuum is frequently created between the inner and outer walls [42].

2.4.1.2 Material requirements

Metal Dewars are preferred for their mechanical strength and gas tightness, which are crucial for high-temperature superconducting (HTS) applications [43].

2.4.1.3 Innovations in Dewar technology

2.4.1.3.1 Quench detection

Advanced Dewar designs incorporate features for quench detection, such as optical fiber sensors, enhancing the safety and reliability of HTS magnets [44].

2.4.1.3.2 Magnetic shielding

Some Dewars include shielding covers to protect electronic components from strong magnetic fields generated by superconducting coils, ensuring stable operation [45].

2.5 Magnet stabilization

Stable magnet operation is not ensured by using stable composite wires and cables in a magnet coil. Numerous disruptions can affect superconducting magnet coils, including conductor displacements during excitation, epoxy cracking, AC losses, splice heating, beam-induced heat deposition, variations in the cryogenic system's helium temperature, etc. They could raise the coil temperature above both the superconductor critical temperature T_C and the current sharing temperature T_{Cs}. This would cause all transport current to be ejected into the matrix and then to begin sharing between the superconductor and the normal matrix. According to M.N. Wilson, Superconducting Magnets, Oxford University Press, New York, 1983, these disturbances vary in size (from small, comparable with the superconducting filament size, to large, comparable with the whole conductor length), time scale (from short heat pulses to steady heat depositions), and coil location. Ensuring that a magnet coil can remain superconducting or regain superconductivity following a disturbance is the primary objective of stability. One major challenge in the design of superconducting magnets is the suppression of instability in the case of quench. To that end, two strategies have been developed: the first involves eliminating flux hopping, while the second limits its effects by cryogenic stabilization. In actuality, one typically combines these two methods [46].

3. Application

3.1 Superconducting magnetic energy storage (SMES)

Electric power is efficiently and quickly extracted from the magnetic field of a massive superconducting magnet in a SMES system. When the SMES is being discharged and when it is being recharged, power conditioning equipment is needed to convert the DC power in the magnet to AC power for the grid. SMES has a number of possible uses in the energy sector. Large units with a capacity of more than 1 GW-h could be utilized for load leveling and diurnal storage. Numerous operational advantages, such as spinning reserve, autonomous generation management, black start capability, and increased system stability, may be offered by smaller units [47].

Because SMEs can supply massive amounts of pulsed power to weapon systems like ground-based lasers for ballistic missile defense, it is also of interest to the military. The development of an engineering test model (ETM) with a capacity of 20 MW-h/400 MW is currently being supported by the Strategic Defense Initiative Organization (SDIO) and might start as early as 1993. Utilities are contributing a small portion of the funding through the Electric Power Research Institute because military and utility designs are similar except for the power conditioning system (weapons need to receive large amounts of power quickly and may drain the SMES in a short time, while utilities need to have a constant, reliable supply from which smaller amounts of power are withdrawn on a daily basis). SMES can be used in a variety of fields to offset the weaknesses in each sector by combining its benefits with those of the other fields. Electric energy can be produced from various forms of energy using a variety of power generation techniques. After that, consumers can use the electricity that has flown to the grid *via* the cable, while SMES stores any electricity that has not yet been consumed. In order to compensate for the lack of power generation during the low power generation phase, electric energy is discharged. This keeps the grid well-supplied while cutting down on the wastage of created extra electricity [47].

3.2 Application of power generation field

3.2.1 Photovoltaic power generation (PVPG)

The method known as photovoltaic power generation uses the semiconductor interface's photovoltaic effect to directly transform light energy into electric energy. Three components make up its fundamental structure: the inverter, controller, and solar panel. SMES can function benignly in PVPG systems in addition to being able to store energy. Over the past few years, our power system has faced numerous issues, including reactive power, unbalanced current, unbalanced voltage, sags and swells in voltage, power variations, load fluctuations, and compensation of induced harmonics [48]. These difficulties are caused by the rapidly expanding use of switched-mode power supplies (SMPS), nonlinear loads, static power converters, ever-expanding and congested power lines, and high permeability renewable energy sources (RESs), such as the sun, wind, and other natural resources [49]. Especially among all kinds of RESs, PVPGS with no pollution, no geographical restrictions, and abundant reserves account for an increasing proportion of microgrids [50]. Unfortunately, solar energy is highly erratic and clearly intermittent, which causes significant variations in photovoltaic output power [51]. SMES can smoothly suppress the power fluctuation caused by load mutation, pulse load, and time-varying solar irradiance [52].

3.2.2 Wind power generation

Wind power generation essentially transforms wind energy into mechanical energy, which is subsequently transformed into electrical energy to produce electricity. The idea behind wind power generation is to harness the wind to rotate the blades of a windmill. Subsequently, the apparatus will accelerate its rotational speed to encourage the generator to produce electricity. The two main issues that modern wind power generation systems (WPGS) face is (1) wind speed has a significant impact on the output power of wind power generation, and this output power fluctuates randomly; and (2) if wind power generation equipment is simply turned off when the power grid voltage drops, this will seriously affect the stability of the power grid. SMES is experimenting with varying wind speeds to manage the voltage of electricity networks integrated with wind farms. The Simpower system package and the MATLAB/Simulink program were used to model SMES, WPGS, and the research system [51]. Additionally, they derive the following conclusions from this experiment to support the benefits of SMES when used in conjunction with WPGS: (1) When WPGS and SMES are put together, the voltage value stays constant regardless of variations in wind speed, suggesting that SMES and WPGS can work together to stabilize the voltage. Reactive power and electrical energy loss in the line can both be decreased *via* SMES [53]. As a result, SMES is crucial to the stability of the wind power generation system. In the meantime, SMES and WPGS work together to suppress externally generated changes in voltage and frequency, much like PVPGS does. The wind speed should be the external element in this integrated system. As a result, SMES-equipped wind power plants can run more steadily [53].

3.2.3 Thermal power generation

Burning combustibles to convert heat energy into electrical power *via* a power generation device is known as thermal power generation. Thermal power generation systems are typically built with Automatic Gain Control (AGC), which is a crucial component of the Energy Management System (EMS). AGC regulates the frequency modulation unit's output to meet users' fluctuating power demands and bring the system into a financially viable operating state [54].

When the system is exposed to variations in load, the AGC's job is to maintain the system frequency and actual power between various control zones at their respective nominal values. The role of AGC in the thermal power system will be strengthened if we integrate it with SMES and AGC. An experiment conducted by Sabita Chaine and M. Tripathy demonstrated the superiority of integrating SMES with thermal energy generation. When used with appropriately calibrated controllers, tachyon energy storage devices like SMES are successful at suppressing oscillations in the power system due to their sensitivity and robustness under a variety of operating situations [54].

3.3 Accelerator magnets

The most crucial instruments in high-energy physics research, accelerators enable the study of the biggest masses and the tiniest spatial scales. The machine radius r (in meters) and the strength of the bending dipole magnets B (in Tesla) set a limit on the energy E (in GeV) of particles in a circular accelerator.

$$E = 0.3rB \tag{1}$$

Therefore, the most effective route to higher energy accelerators is *via* high fields. A superconducting accelerator's magnet system is made up of quadrupole magnets that are employed as lenses to focus particle beams and bending dipole magnets that maintain particles on a closed orbit. To adjust beam orbit and parameters, specific dipole, quadrupole, sextupole, and octupole magnets are also utilized [55]. The ring circumference determines the accelerator construction cost, while the magnet power consumption determines the accelerator operation cost. Thus, in order to reduce coil volume and magnet cost, superconducting magnets for particle accelerators depend on the maximum current density. The other important factors for accelerator magnets are field quality and its repeatability from magnet to magnet. Generally, they need to be at least 0.01% of the primary field component.

3.4 Superconducting magnet for MRI

Nuclear magnetic resonance, in which an atomic nucleus put in a magnetic field absorbs the energy of radio waves at a given frequency, is the basis for magnetic resonance imaging (MRI). Using primary magnets and radio frequency coils, MRI creates images of the distribution of protons, or hydrogen nuclei, throughout the human body. As a result, using noninvasive external magnetic fields to detect and analyze diseases early and more accurately has made it an essential diagnostic tool. As a result of ongoing technological advancements, MRI systems have been developed that are increasingly useful for neurosurgical planning, angiography, and cancer detection. The field strength and imaging resolution are inversely related. MRI magnets that are currently in widespread use in clinical settings have a magnetic field intensity of 1.5 T. Analysis of the body compositions of much smaller forms and observation of finer body structures are made possible when this strength is much boosted. Since then, magnetic resonance imaging (MRI) has spread throughout the globe as a vital medical tool for *in vivo* imaging. This makes MRI maybe the greatest invention since X-ray machines, which provided the first diagnostic view inside the human body. With high-resolution, soft-tissue contrast, functional process and imaging, and a safe 3D image of the human body, magnetic resonance imaging (MRI) replaces X-rays, which pose a risk to human cells [56].

3.5 The SQUID magnetometer

When a weak link separates two superconductors, a Josephson junction is formed. Such a shaky connection can result from several factors. A SQUID magnetometer consists of one or two of these connections in a superconducting loop. The DC-SQUID with two Josephson junctions is used to detect biomagnetic signals in most applications.

To maintain a low inductance, the SQUID loop should be as small as possible; a pickup loop is necessary to have a small sensor loop with a high level of sensitivity. Some high-temperature superconductors use a directly connected pickup loop, however, this loop typically links inductively with the SQUID loop. Together with the sensor, certain readout electronics are also needed. Current locked loops and fluxed locked loops are the most commonly used reading systems, while there are others [57].

3.5.1 Magnetoencephalography (MEG)

Biomagnetic signals are primarily produced by intracellular current flow, which is triggered by stimulation of muscles or nerves. These cells act as the current source, and the circuit is completed by the extracellular current passing through the volume conductor. This current flow produces quantifiable electric fields, which can be measured using electroencephalography (EEG) techniques. Additionally, a very faint but discernible magnetic field is produced. One advantage of measuring the magnetic field is that it makes it possible to measure currents that are perpendicular to the surface of the body. EEG can attain far lower resolutions than MEG because volume current is more distorted than intracellular current flow. Intracellular current flow is the primary cause of these currents. Additionally, because the two metrics complement each other, it can be used in conjunction with EEG approaches [58].

MEG systems typically have a Dewar around the entire head, including a range of low-temperature SQUID sensors (see **Figure 4**). Typically, the array is made up of 37–255 SQUID sensors arranged either as a magnetometer or as a gradiometer, with a small number of sensors devoted to noise cancelation. Two pickup coils placed on a baseline in a gradiometer arrangement are arranged so that the magnetic fields flowing through them cancel out and only the difference is amplified (**Figure 5**). The magnetometer consists of a single SQUID sensor and pickup coil. MEG measurements are always conducted in a magnetically protected chamber due to the small signal sizes.

Numerous clinical and research applications employ MEG. While MEG is an excellent tool for certain tasks, such as functional mapping of cortical areas, it is also a unique tool for other tasks, such as examining the temporal aspects of the brain's signal processing and other even more complicated brain activities. When combined with EEG, functional magnetic resonance imaging (fMRI), positron emission tomography (PET), and MRI, MEG offers a variety of additional special study applications.

The primary use of MEG in clinical settings is to locate lesions and epileptic discharge centers. Using this in conjunction with MRI or fMRI allows for precise preoperative planning. Other uses include the planning of stereotactic radiation therapy by precisely localizing the primary sensory cortex through functional mapping, which keeps radiation to this region as low as possible, and the planning of rehabilitation for stroke patients by assessing the extent of brain damage [59].

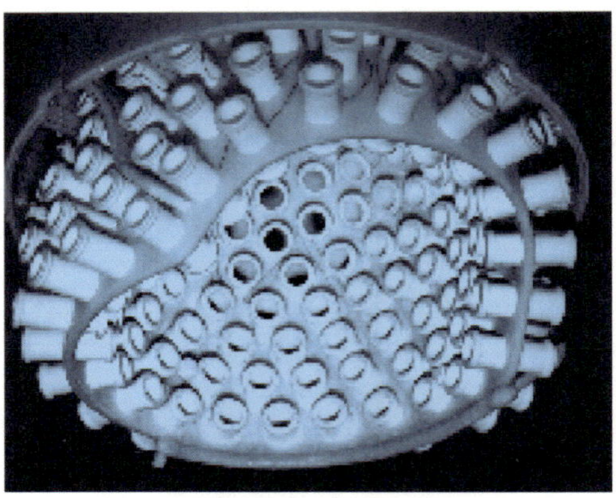

Figure 4.
Whole head MEG system.

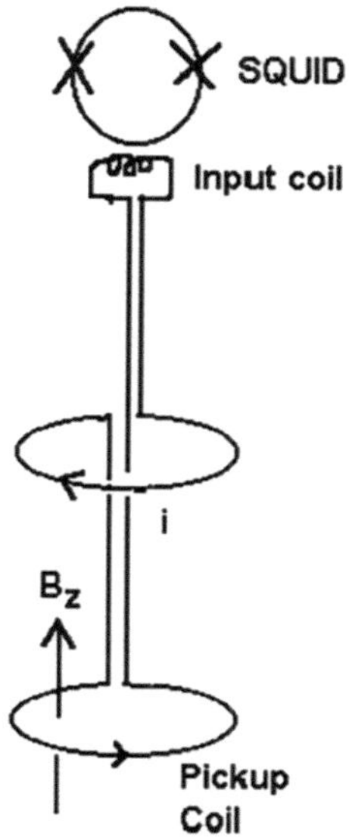

Figure 5.
Gradiometer configuration.

3.5.2 Magnetocardiography (MCG)

The magnetic field produced by the heart's current flow is about 50 pT in strength. Both low and high-temperature SQUIDS can detect this well within their detection range. This is the first area of medicine where using high-temperature SQUIDs is now a practical substitute. Once more, the intracellular current flow is measured by the magnetometer and is considerably less distorted by the body than the volume current flow detected by EEG devices. Another completely noninvasive technique that does not need electrodes is MCG. The primary barrier to MCG is the requirement for shielding; before the clinical application becomes a reality, a few unshielded sensors still need to be enhanced. An array of SQUID sensors, which frequently cover the entire chest region and occasionally the front and back as well, is another feature of the basic MCG device. There are between seven to 64 sensors. Every experimental model, with the exception of a handful, works in a magnetically insulated chamber [58].

Since MCG preparation takes far less time than ECG measurements, it is useful in-patient screening. It has also been utilized in conjunction with the catheter technique to map cardiac arrhythmias. The MCG is used to map the catheter position in order to obtain a more accurate assessment of the area that has to be ablated once the arrhythmia has been broadly localized. In addition to being used to identify myocardial ischemia, MCG has been demonstrated to predict malignant cardiac arrhythmias with significantly higher accuracy than ECG [59]. Fetal magnetocardiography is one of the primary uses of MCG that is currently emerging. Because the vernix caseosa, an isolating layer in fetal ECGs, is prone to severe distortion, MCGs have great clinical value for analyzing fetal heartbeats in high-risk pregnancies because they are not subject to this distortion [56].

3.5.3 Magnetoneurography

Magnetoneurography is the measurement of peripheral nerve impulses using magnetic fields. Peripheral nerve signals are so small—about 15 fT—that they can only be detected by highly sensitive, low-temperature SQUIDs in severely shielded rooms. Even then, special noise cancelation is required because the magnetic disturbance caused by the heart is nearly a 1000 times larger [60]. This method is useful for locating conduction blocks or slowing in peripheral nerves, and it may be used as a diagnostic procedure in this regard. In situations where there is an acute nerve root lesion along with excruciating back pain and spasms in the muscles, it can also be used to find the focal nerve [61].

3.6 Power application

Most applications in electric power equipment require material in a wire or tape configuration, with only a few exceptions. It has been suggested that superconducting materials in the form of "bulk" blocks, plates, or rods be used in non-wire arrangements for devices like power leads, flywheel bearings, and fault current limiters. All practical applications of superconducting wire, coils, and magnets inherently result in significant energy savings because superconducting wire provides zero resistance to the flow of direct electrical current. True technological advancements have been made possible by the creation of strong magnetic fields in a workable working volume with the least amount of energy needed, mostly for cooling. The compactness of size (MRI), low weight (MAGLEV), high specific performance (NMR), and other characteristics of superconducting systems provide the customer considerable advantages for non-energy applications of bulk superconducting material. Energy reductions alone do not justify the use of superconducting technology in energy-related applications. Superconductors must be able to demonstrate their performance, dependability, safety, and lower total life cycle cost before they can be integrated into the electrical grid.

Superconductors do not impede the flow of electric current while they are operating in a steady state of direct current. On the other hand, superconductors do not have zero resistance and do show some losses in AC or transient situations. They were therefore initially used in dc or quasi-DC environments. Low ac loss conductors have been created especially for operation under transient or steady state AC operation as the loss mechanism of conductors at power frequencies has been better understood. In essence, full-rating developmental devices have been constructed in a number of instances to show how devices like rotating synchronous generators and AC power transmission lines operate. These devices are typically constructed using LTS NbTi [62].

Since the construction of an 8 kW 400 Hz demonstration machine by Stekly et al. in the 1960s, superconductors have been used in rotating synchronous generators [63]. This generator used a spinning room temperature armature and a stationary superconducting field winding because the existing superconductors could only be used in direct current (DC) applications. Subsequent machines were designed using a stationary room temperature armature and a spinning superconducting field winding as the preferred form. A 20 MVA unit was constructed and tested by General Electric in the early 1980s, operating at full load output power [64].

3.6.1 Transformers

Transformers with superconducting windings have been the target of multiple efforts since the 1960s when superconductors became economically viable for use in large magnets. A group at Westinghouse conducted a comprehensive feasibility assessment of superconducting windings in a big power transformer more than 10 years ago. Despite the transformer's superior qualities resulting from the use of known or deemed developable superconductors, the team discovered that as long as the windings remained electrically connected to the network after a fault, they would not be able to return to the superconducting state [63].

3.7 Superconducting magnets in transportation (maglev)

The term "maglev" comes from "magnetic levitation," which is the process of lifting a vehicle using an electromagnet or permanent magnet to create a magnetic field and eliminate the need for any mechanical support. Another name for superconducting maglevs is linear motor vehicles. The motor is not rotary; it is linear. It can be compared to a regular electric motor that has been divided open, laid out flat, and pointed in the direction that a train is moving. The maglev train technology uses two sets of magnets.

Two sets of magnets are used: one for levitating the train body off the track, and another for propulsion using attraction and repulsion. The moving body of a levitated train has additional stability and directional control thanks to guideways in the form of magnets. It is prevented from moving outside of its magnetic field range by the magnetic field these guide way magnets produce. Because of this, the maglev train is more reliable and efficient than the conventional train. Because there are no moving parts and no direct touch between the train body and the tracks, the operation is frictionless [64]. The magnetic field produced by electrified coils powers the maglev train instead of an engine.

There are three primary types of maglev technologies. Three main categories of maglev technologies exist because first to raise the train, electromagnetic suspension (EMS) uses an attracting magnet beneath a rail to generate a magnetic force. Under this arrangement, the train is drawn to a magnetically conductive (often steel) track by electronically controlled electromagnets within the vehicle. Second, to push the train away from the rail, electrodynamic suspension (EDS) applies a repulsive force between two magnetic fields. Strong permanent magnets or superconducting electromagnets in this system produce a magnetic field that, when there is a relative movement, produces currents in adjacent magnetic conductors, pushing and pulling the train in the direction of the intended levitation location on the guide path. Third, lift the train above the ground, stabilized permanent magnet suspension (SPM) uses opposing arrays of permanent magnets above the rail. The magneto dynamic suspension (MDS) is another experimental device that was created, mathematically demonstrated, peer-reviewed, and patented but has not yet been put into use. It works by using the magnetic pull of a permanent magnet array placed close to a steel track to draw in trains and keep them there [65].

The three primary functions of magnetic levitation are: (1) to lift the vehicle, (2) to provide propulsion and attain a rapid pace, and (3) to offer direction. The working principle of the maglev system depends on: (1) the magnetic field generated through permanent magnets or electromagnets and (2) the magnetic poles of magnets or electromagnets. Levitation to vehicle or object due to magnetic field generated by respective power source, whether it be permanent magnets or electromagnets if the same pole of respective power source faces to each other and propulsion of vehicle due to attraction or repulsion action of the magnetic field generated by another same power source as their opposite poles face to each other is the working principle of a maglev train (**Figure 6**) [66].

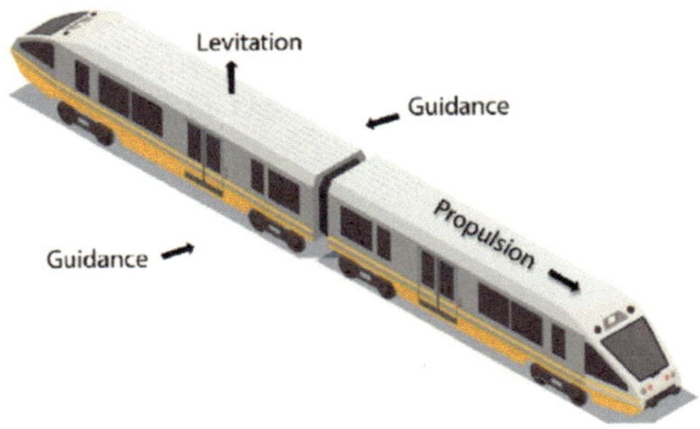

Figure 6.
Primary function of maglev [66].

4. Conclusion

Superconducting magnets have revolutionized various fields, from energy generation and storage to medical diagnostics and transportation and also scientific research. Their ability to operate at extremely low temperatures and generate powerful magnetic fields has enabled numerous advancements. Their ability to generate powerful magnetic fields with high efficiency and minimal energy consumption has led to significant advancements in technology and scientific research. As research and development continue, we can expect even more innovative applications of superconducting magnets in the future.

Acknowledgements

We acknowledge our wives for their unlimited support.

Funding

The authors did not receive any financial support for this manuscript.

Conflict of interest

The authors declare that there is no conflict of interest.

DOI: http://dx.doi.org/10.5772/intechopen.1007576

Author details

Habtamu Anagaw[1,2*] and Gedefaw Mebratie[1]

1 Department of Physics, College of Science, Bahir Dar University, Bahir Dar, Ethiopia

2 Department of Physics, College of Science, Dire Dawa University, Dire Dawa, Ethiopia

*Address all correspondence to: habtana@gmail.com

References

[1] Kamerlingh Onnes H. On the sudden change in the rate at which the resistance of mercury disappears, in KNAW. Proceedings. 1911;**14**:818

[2] London F, London H. The electromagnetic equations of the superconductor. Proceedings of the Royal Society A, Mathematical, Physical and Engineering Sciences. 1935;**149**:71

[3] Annett JA. Superconductivity, Superfluids and Condensates. Oxford University Press; 2004

[4] Bardeen J, Cooper LN, Schrieffer JR. Theory of superconductivity. Physics Review. 1957;**108**(5):1175

[5] Boeri L et al. The 2021 room-temperature superconductivity roadmap. Journal of Physics: Condensed Matter. 2022;**34**:183002 (51 pp)

[6] Frolich H. Theory of the superconducting state. I. The ground state at the absolute zero of temperature. Physics Review. 1950;**79**:845

[7] Rourke PMC. Electronic states of heavy fermion metals in high magnetic fields, A thesis submitted to University of Toronto. 2009

[8] Ott HR, Rudigier H, Fisk Z, Smith JL. UBe_{13}, an unconventional actinide superconductor. Physical Review Letters. 1983;**50**(20):1595

[9] Whitea BD, Thompsonc JD, Maple MB. Superconductivity and its applications. Physica C. 2015;**514**:246

[10] Werthamer NR, Helfand E, Hohenberg PC. Temperature and purity dependence of the superconducting critical field, H_{C2}. III electron spin and spin-orbit effects. Physics Review. 1966;**147**:295

[11] Wu T, Mayaffre H, Krämer S, Horvatić M, Berthier C, Hardy WN, et al. Magnetic-field-induced charge-stripe order in the high-temperature superconductor $YBa_2Cu_3O_y$. Nature. 2011;**477**:191

[12] Lake B, Rønnow HM, Christensen NB, Aeppli G, Lefmann K, McMorrow DF, et al. Antiferromagnetic order induced by an applied magnetic field in a high-temperature superconductor. Nature. 2002;**415**:29901

[13] Kamihara Y, Watanabe T, Hirano M, Hoson H. Iron-based layered superconductor $La[O_{1-x}F_x]FeAs$ (x = 0.05-0.12) with T_C = 26 K. Journal of the American Chemical Society. 2008;**130**:3296

[14] Shishido H, Bangura AF, Coldea AI, Tonegawa S, Hashimoto K, Kasahara S, et al. Evolution of the Fermi surface of BaFe2 $(As_{1-x}P_x)_2$ on entering the superconductingdome. Physical Review Letters. 2010;**104**:057008

[15] Larbalestier D, Gurevich A, Feldmann DM, Polyanskii A. High-T C superconducting materials for electric power applications. Nature. 2001;**414**:368

[16] Lebed A. The Physics of Organic Superconductors and Conductors, Springer Series in Materials Science 110. Berlin, Heidelberg, New York: Springer-Verlag; 2008

[17] Rogalla H, Kes PH. 100 Years of Superconductivity. Boca Raton, Florida: Taylor and Fransis Group, LLC; 2012

[18] Ashby M, Shercliff H, Cebon D. Materials (4th Edition), Engineering, Science, Processing and Design. London,

United Kingdom: ICE Publishing; 2019. pp. 447-474. DOI: 10.1016/B978-0-08-102376-1.00016-0

[19] Iwasa Y. Case Studies in Superconducting Magnets: Design and Operational Issues. Boston: Springer Science + Business Media, LLC; 2009. DOI: 10.1007/b112047_1

[20] Xiaojing Z. Preface: Mechanical perturbations induced quench: A challenge of superconductor mechanics. National Science Review. 2023;**10**:1-2, nwad044. DOI: 10.1093/nsr/nwad044

[21] Tong Y, Guan M, Wang X. Theoretical estimation of quench occurrence and propagation based on generalized thermoelasticity for LTS/HTS tapes triggered by a spot heater. Superconductor Science and Technology. 2017;**30**:045002. DOI: 10.1088/1361-6668/aa595a

[22] Tsukamoto O, Maguire J, Bobrov E, Iwasa Y. Identification of quench origins in a superconductor with acoustic emission and voltage measurements. Applied Physics Letters. 1981;**39**:172-174

[23] Ninomiya A, Sakaniwa K, Kado H, Ishigohka T, Higo Y. Quench detection of superconducting magnets using ultrasonic wave. IEEE Transactions on Magnetics. 1989;**25**:1520-1523

[24] Dixon IR, Gavrilin AV, Miller JR, Powell JA, Brandt BL. Analysis of the quench protection system of a series connected hybrid magnet. IEEE Transactions on Applied Superconductivity. 2007;**17**:2446-2449

[25] Wang X, Guan M, Ma L. Strain-based quench detection for a solenoid superconducting magnet. Superconductor Science and Technology. 2012;**25**:095009

[26] Scurti F, Ishmael S, Flanagan G, Schwartz J. Quench detection for high temperature superconductor magnets: A novel technique based on Rayleigh-backscattering interrogated optical fibers. Superconductor Science and Technology. 2016;**29**:03LT01

[27] Schwartz J. Quench in high temperature superconductor magnets. In: CERN Yellow Report CERN-2013-006. Geneva, Switzerland: CERN; 2014. pp. 21-29. DOI: 10.5170/CERN-2013-006.21

[28] Yamada R, Marcsin E, Lee A, Wake M. 3D ANSYS quench simulation of cosine theta Nb_3Sn high field dipole magnets. IEEE Transactions on Applied Superconductivity. 2004;**14**:291-294

[29] Kalsi SS. Applications of high temperature superconductors to electric power equipment. In: IEEE Press. Piscataway, NJ: A John Wiley & Sons, Inc.; 2011. pp. 36-41

[30] Cavallucci L, Gauthier F, Breschi M, Hoa C, Bauer P, Vostner A. Multiphysics model of quench for the ITER central solenoid. IEEE Transactions on Applied Superconductivity. 2023;**33**(5):1-5

[31] Iwasa Y. Case Studies in Superconducting Magnets: Design and Operational Issues. New York: Plenum Press; 1994

[32] Wilson MN. Superconducting Magnets. Oxford: Clarendon Press; 1983

[33] Zhanguo Z, Norihito O, Masanori K, Aoki K, Toshiyuki O, Xudong W, et al. Cryogenic systems of SuperKEKB final focusing superconducting magnets. Nuclear Instruments and Methods in Physics Research. 2024;**1058**:168855. DOI: 10.1016/j.nima.2023.168855

[34] Xue R et al. A long-life, high-capacity and high-efficiency cryogenic system developed for high-T_C superconducting magnet applications. IEEE Transactions on Applied Superconductivity. 2022;**32**(6):1-5. DOI: 10.1109/TASC.2022.3153234

[35] Shoichi Y. Technical overview of superconducting magnets. Japanese Society for Magnetic Resonance in Medicine. 2023;**43**(1):1-10. DOI: 10.2463/jjmrm.2022-1771

[36] Yang W, Wu B, Ni D, Wu W, Mei E, Lizhen M. Cryogenic test of the frib superconducting magnets at IMP. Journal of Physics: Conference Series. 2020;**1401**:012017. DOI: 10.1088/1742-6596/1401/1/012017

[37] Thekkethil et al. Effect of thermal strain, induced by cryogenic cooling, on a high homogeneity superconducting magnet for MRI applications. Indian Journal of Engineering & Materials Sciences. 2022;**29**:581-585. DOI: 10.56042/ijems.v29i5.52987

[38] Hang Z, Hansheng F, Rui H. Simulation and structure design of high homogeneity conduction cooled superconducting magnet. In: IEEE International Conference on Applied Superconductivity and Electromagnetic Devices (ASEMD). Tianjin, China: IEEE; 2020. pp. 1-2. DOI: 10.1109/ASEMD49065.2020.9276148

[39] Wang Y, Jing Z. Multiscale modeling on the multi-physical behaviors of high temperature superconducting magnets based on a combined global homogenization and local refinement scheme. Superconductor Science and Technology. 2023;**313**:116863. DOI: 10.1088/1361-6668/ad54f7

[40] Łukasz T, Kulikov E, Kozłowski K, Drobin V. Improvement of the homogeneity of magnetic field by the attenuation of a selected component with an open superconducting shield made of commercial tapes. Journal of Applied Physics. 2019;**126**:083903. DOI: 10.1063/1.5112036

[41] McIntyre P. Integration and testing of the superconducting magnet and cryogenics for AMS. IEEE Transactions on Applied Superconductivity. 2010;**20**(3):2010-2014. DOI: 10.1109/TASC.2010.2040029

[42] Dong F, Hao L, Park D, Iwasa Y, Huang Z. On the future sustainable ultra-high-speed maglev: An energy-economical superconducting linear thrusting system. Energy Conversion and Management. 2023;**291**:117247

[43] Liu G, Gao R, Tian C, Wang D, Wu Y, Wang Y, et al. Study of a low-frequency wireless charging system crossing the metal Dewar wall of a high-temperature superconducting magnet. Superconductor Science and Technology. 2023;**36**:015005. DOI: 10.1088/1361-6668/aca368

[44] Liu G, Gao R, Tian C, Wang D, Wu Y, Wang Y, et al. Study of a low-frequency wireless charging system crossing the metal Dewar wall of a high-temperature superconducting magnet. Superconductor Science and Technology. 2022;**36**(1):015005

[45] Chen Z, He P, Li Y, Yang X, Wei J, Zhang X, et al. Magnetic field measurement system in vertical status for superconducting undulator cooled in Dewar. Journal of Instrumentation. 2022;**17**(9):T09008

[46] Wilson MN. Superconducting Magnets. New York: Oxford University Press; 1983

[47] Schauer F et al. Assessment of Potential Advantages of High T_C Superconductors for Technical Application of Superconductivity. Karlsruhe, Germany: Kernforschungszentrum Karlsruhe. Kfk4308; 1987

[48] Bollen MH, Das R, Djokic S, Ciufo P, Meyer J, Rönnberg SK, et al. Power quality concerns in implementing smart distribution-grid applications. IEEE Transactions on Smart Grid. 2016;**8**(1):391-399

[49] Tareen WU, Mekhilef S, Seyedmahmoudian M, Horan B. Active power filter (APF) for mitigation of power quality issues in grid integration of wind and photovoltaic energy conversion system. Renewable and Sustainable Energy Reviews. 2017;**70**:635-655

[50] Bialasiewicz JT. Renewable energy systems with photovoltaic power generators: Operation and modeling. IEEE Transactions on Industrial Electronics. 2008;**55**(7):2752-2758

[51] Justo JJ, Mwasilu F, Lee J, Jung J-W. AC-microgrids versus DC-microgrids with distributed energy resources: A review. Renewable and Sustainable Energy Reviews. 2013;**24**:387-405

[52] Jin JX, Wang J, Yang RH, Zhang TL, Mu S, Fan YJ, et al. A superconducting magnetic energy storage with dual functions of active filtering and power fluctuation suppression for photovoltaic microgrid. Journal of Energy Storage. 2021;**38**:102508

[53] Said SM, Aly MM, Hartmann B. Application of SMES for voltage control of power systems with high wind power penetration. In: 2018 International Conference on Innovative Trends in Computer Engineering (ITCE). Aswan, Egypt: IEEE; 2018. pp. 461-466

[54] Chaine S, Tripathy M. Design of an optimal SMES for automatic generation control of two-area thermal power system using Cuckoo search algorithm. Journal of Electrical Systems and Information Technology. 2015;**2**(1):1-13

[55] Mess K-H, Schmuser P, Wolff S. Superconducting Accelerator Magnets. Singapore: World Scientific; 1996

[56] Cosmus T, Parizh M. Advances in whole-body MRI magnets. IEEE Transactions on Applied Superconductivity. 2010;**21**:2104-2109

[57] Conradie EH. The design and fabrication of DC SQUID magnetometers [masters thesis]. Stellenbosch, South Africa: Electronic Engineering, University of Stellenbosch; 1998

[58] Andra W, Nowak H. Magnetism in Medicine: A Handbook. Weinheim, Germany: Wiley-VCH; 1998. pp. 176-178

[59] Biomag. Book of abstracts. In: 12th International Conference on Biomagnetism; August 13-17. Espoo, Finland: Helsinki University of Technology; 2000, 2000

[60] Burghoff M, Mackert B-M, Haberkorn W, Curio G, Trahms L. Highresolution magnetoneurography. Applied Superconductivity. 1998;**6**(10-12):567-575

[61] Mackert B-M, Curio G, Burghoff M, Thrahms L, Marx P. Magnetoneurographic 3D localization of conduction blocks in patients with unilateral S1 root compression. Electroencephalography and Clinical Neurophysiology. 1998;**109**:315-320

[62] Sokolowski RS, Rosner CH. Industrial Applications of Superconductivity. Monterey, CA, USA: American Institute of Aeronautics

and Astronautics, Inc.; 1994.
DOI: 10.2514/6.1994-4216

[63] Stekly ZJJ, Woodson HH, Hatch AM, Hoppie LO, Halas E. A study of alternators with superconducting field windings: II-experiment. IEEE Transactions on Power Apparatus and Systems. 1966;**85**(3):274-280

[64] Dhingra M, Sharma R, Salmani M. An introduction and overview to magnetically levitated train. Journal of Science. 2015;**5**(11):1117-1124

[65] Vignesh M, Vigneshwaran R, Vijay V. Design of magnetic levitation train. International Journal of Innovative Research in Science, Engineering and Technology. 2015;**4**(2):2347-6710

[66] Tambe AT, Patil DS. Introduction & overview of magnetic levitation (MAGLEV) train system. International Journal of Innovative Science and Research Technology. 2018;**3**(2):2456-2165

Section 4

Emerging Technologies

Chapter 7

Latest Development of Superconducting Maglev

Wei Zhou, Kun Liu, Zhihua Zhang and Si Yuan Liang

Abstract

The suggested chapter will present the latest developments of the superconducting maglev in the world (especially in China). The chapter firstly introduces *the classification of suspension*, for example, the developments of electro-magnetic suspension (EMS), electro-dynamic suspension (EDS), and magnetic flux pinning suspension. Secondly, *we explain the concept and principle of EMS, EDS*, and so on. Thirdly, we *give the introduction to the latest developments of the maglev in the world*, such as China, Japan, the US, and so on. In particular, the chapter will give more details about the superconducting EDS developments in China. Finally, the chapter will *present the key technologies and* research prospect *of the superconducting magnet of the EDS*.

Keywords: maglev, high speed, electro-dynamic suspension, superconducting magnet, key technologies

1. Introduction

With the development of human society and economy and the progress of science and technology, people's demand for transportation capacity is increasing day by day, and there is an urgent need for a faster and more efficient transportation system. In the future transportation field, the application of maglev technology undoubtedly has more advantages. There are many kinds of magnetic levitation systems, such as electrical-dynamic levitation (EDS), electromagnetic suspension (EMS), and superconducting pin levitation. EDS has become one of the fastest developing systems because of its advantages of large suspension gap and simple control logic. Superconducting magnet is the "magnetic source" of EDS as well as the power source of the whole system. Japan, the United States, and other countries carried out research on EDS superconducting magnet as early as the 1970s and successively completed the principal test verification, engineering test verification, and commercial test operation. This paper focuses on the latest research progress of EDS, especially the development of EDS in China. Finally, key technologies and research prospect of superconducting magnets for EDS are introduced.

2. The classification of suspension

Maglev trains reduce the friction resistance between the wheels and the track, and the maximum speed can reach 600 km/h or even higher, which has great potential in the future transportation system. According to the different suspension modes of maglev trains, there are currently three types of maglev trains: Electromagnetic maglev train (EMS), electrical-dynamic maglev train (EDS), and superconducting pin suspension train (SPL) [1]. The characteristics and related parameters of the three maglev technologies are shown in **Table 1**.

Among the above three levitation systems, EMS and EDS have reached the practical level. Among them, EDS has become the mainstream development trend of ultra-high-speed maglev due to its high running speed, light vehicle, and large suspension gap. In the EDS system, the EDS train developed in Japan has the most practical prospect. The basic composition of the system is shown in **Figure 1**. The system adopts "U" shape track, and the superconducting magnet is suspended on both sides of the car body, and the "8" suspension guidance coil and propulsion coil are installed on both sides of the track, opposite to the superconducting magnet. The driving of the train relies on the interaction force between the propulsion coils on both sides of the track and the different magnetic poles of the superconducting magnet. The AC current frequency in the propulsion coil is constantly adjusted by the change switch to match the speed of the train to complete the forward and stop of the train, as shown in **Figure 2**. When the train is running, the "8" coil on both sides of the track will generate induced current and magnetic field due to the passing of the superconducting magnet, and the induced magnetic field will react to the superconducting magnet to generate levitation force

Suspension system	**EMS**	**EDS**	**SPL**
Application cases			
Magnet mounted	Constant conductance coil	Superconducting magnet	HTS bulk
Track	Silicon steel sheet	'8' coils	Permanent magnet
Suspension gap	8-12 mm	80-150 mm	10-30 mm
Technique feature	Mature technology, energy consumption, small suspension gap, complex control logic	The lifting and floating speed is required, the superconducting cost is high, the damping is introduced to ensure stability, and the suspension gap is large	No power, no source suspension, self-stable, simple control logic
Application scenarios	Shanghai Maglev line, CRRC maglev	Japan central Shinkansen, aerospace three research institute, Datong test line	Southwest Jiaotong University 165 m test line

Table 1.
Comparison of three maglev systems.

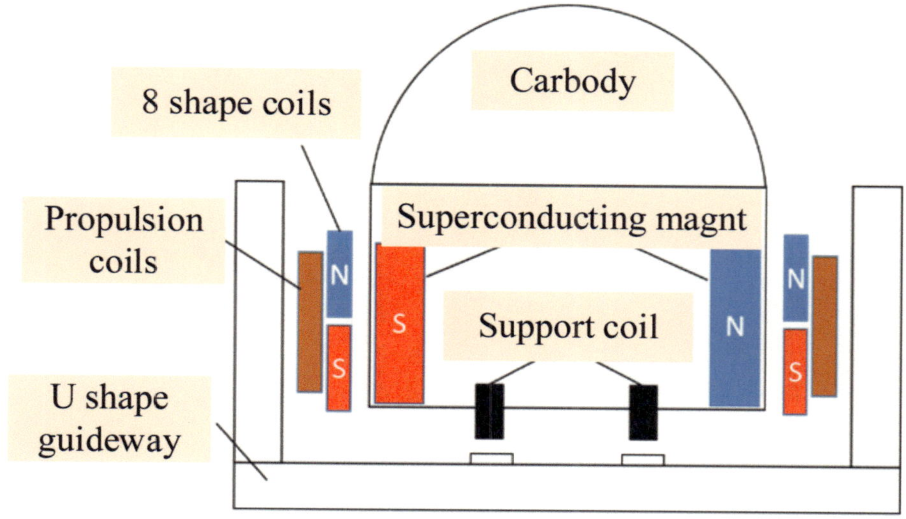

Figure 1.
The basic composition of superconducting EDS train.

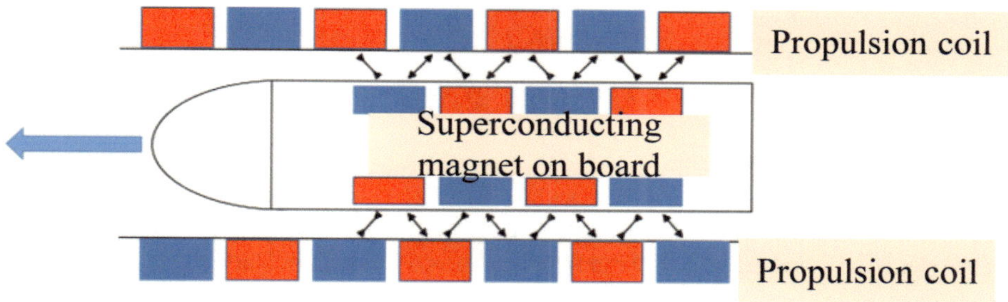

Figure 2.
Train driving principle.

and guiding force. When the speed of the train is low, the levitation force generated by the "8" coil is small, which is not enough to levitate the train. At this time, the train relies on the support wheel to move forward. When the train reaches a certain speed, the levitation force will increase enough to overcome the gravity of the vehicle and achieve levitation, as shown in **Figure 3** [2].

According to the working principle of the superconducting EDS train, it can be seen that the driving power of the train comes from the interaction force between the superconducting magnet on board and the track ground coil. It can be seen that the superconducting magnet on board (hereinafter referred to as the magnet) is equivalent to the "engine" of the maglev train and is the core equipment of the train, so the research on the magnet is very important. The operation of the magnet mainly depends on the zero-resistance characteristic of the superconducting material. When the superconducting coil is below the critical temperature, its resistance is zero. At this time, a certain current is passed into the coil, and the current will flow permanently in the superconducting coil, generating a stable and strong magnetic field.

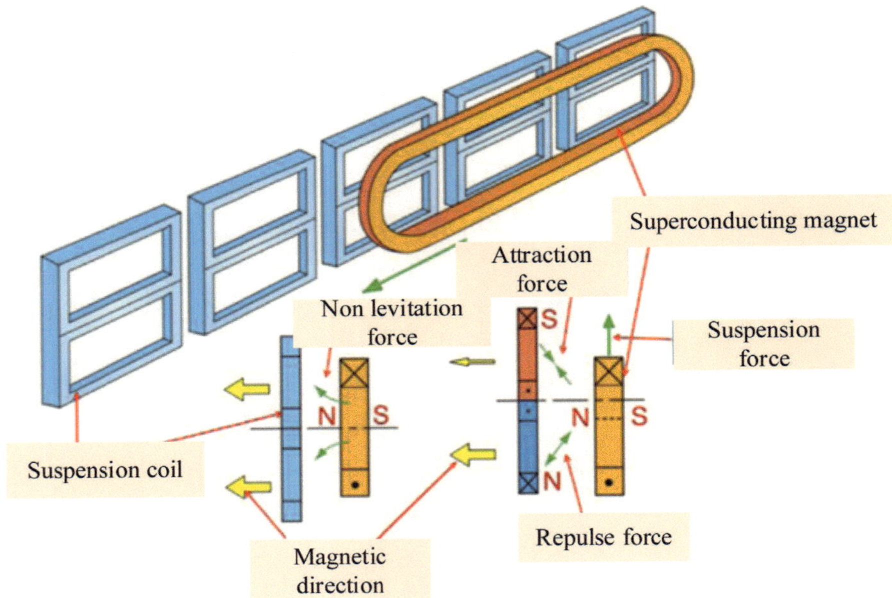

Figure 3.
Schematic diagram of train suspension.

3. The latest developments of the maglev in the world

3.1 Japan central Shinkansen

Japan's Central Shinkansen is a high-speed maglev railway project based on superconducting EDS system, which runs from Tokyo to Osaka via Nagoya and is expected to operate at a maximum speed of 505 kilometers per hour. In 1990, Japan carried out the construction of a pilot line in the Yamanashi prefecture section of the Central Shinkansen, which was completed in 1997. In December of the same year, the high speed of 550 kilometers per hour was achieved, and the speed of 603 kilometers per hour was achieved in 2015. **Figure 4** shows the Japan maglev test line, maglev vehicle, and the low-temperature superconducting magnet.

Throughout the development of Japan's superconducting maglev industry, Japan first began to think about it in the 1970s, completed the basic test inside the research institute in 1972, carried out the navigation experiment of Miyazaki experimental line in 1977, and the navigation experiment of Yamanashi experimental line in 1997. The specific date and vehicle code are shown in **Table 2**. With the development of the vehicle, the superconducting magnet technology is also constantly updated. The low-temperature superconducting magnet and high-temperature superconducting magnet both have carried on the MLX-01. **Table 2** summaries the development of Japan EDS.

3.2 Maglev in USA

In 1969, Massachusetts Institute of Technology started maglev train research [3], using superconducting magnets as suspension and guidance, linear synchronous

(a) (b)

Figure 4.
Japan maglev test line (a) maglev vehicle, (b) LTS magnet.

Serial number	age	Test line	vehicle type	Magnet tyep	Test velocity
1	In 1972	Internal test line	ML100	Horizontal type-LTS	60 km/h
2	In 1977	Miyazaki Test Line	ML500	L type-LTS	517 km/h
3	In 1979		ML500R	L type-LTS	—
4	In 1980		MLU001	I type-LTS	400 km/h
5	In 1986		MLU002	I type-LTS	394 km/h
6	In 1993		MLU002N	I type-LTS	411 km/h
7	In 1997	Yamanashi test line	MLX-01	I type-LTS	552 km/h
8	In 2005		MLX-01	I type-Bi HTS	—
9	In 2015	Central New Trunk Line	L0	I type--LTS	603 km/h

Table 2.
Japan maglev specific and vehicle code.

motor as propulsion and active damping technology; developed a small prototype of 1 m long and 14 kg weight to carry out relevant verification experiments; then carried out research on the plane characteristics of magnetic field, and made experiments with small model. In the same year, North American Rockwell Company theoretically analyzed the feasibility of applying EDS technology to rocket launch. Researchers installed superconducting magnets at the bottom of the rocket sky-car and placed them in the U-shaped track made of aluminum. After the sky-car reached a certain speed, it could achieve stable suspension, so that the rocket could be accelerated to 5000 m/s in the vacuum pipe [4].

In 1974, Ford Aeroneutronics started research and development work on a high-speed maglev train with a test speed of 500 km/h [5, 6]. But with the end of funding under the 1975 High Speed Surface Transportation Act, the US government stopped supporting research on various models of high-speed surface transportation systems. In the late 1980s, the research on maglev trains in the United States heated up again. In 1990, the United States government organized and invested in a National Maglev Initiative; the various performance indicators of three system concept designs of EDS system maglev train proposed by Bechtel, Foster-Miller, and Magne-plane were carefully compared and analyzed, and it was found that the three schemes of superconducting EDS system maglev train had significant differences in train speed and levitation height. It is superior to France's conventional wheel-rail high-speed train TGA-A and Germany's constant conductance electro-maglev high-speed train TR-07 in terms of degree, start time, safety, and comfort.

In the mid-1990s, to upgrade the high-speed test track at Holloman Air Force Base, the United States launched the Holloman Maglev Program (schematic diagram shown in **Figure 5**) [7]. The propulsion system uses direct rocket propulsion, and the levitation system uses the principle of inductive plate superconducting electric magnetic levitation, with the help of the relative motion between the superconducting magnet and the orbital copper plate to generate eddy current to levitate it. In a test in 2003, the high-speed skid system once reached the highest ground running speed of 2884.7 m/s [8–11].

3.3 Maglev in the institute of magnetic levitation and electromagnetic propulsion

The Institute of Magnetic Levitation and Electromagnetic Propulsion, China Aerospace Science and Industry Co. Ltd. carries out the research on EDS of low-temperature superconducting magnets and high-temperature superconducting magnets together. In 2018, a 400-meter test line was built in Hebei province for the proof of principle of electric levitation.

In November 2021, the China Aerospace Science and Industry Magnetoelectricity General Department completed a 623 km/h high-speed flight test on the 400-meter test line [12]. **Figure 6** shows the prototype superconducting vehicle on the guideway of the 400-meter test line. In the test, the vehicle's suspension attitude was stable, the control of the variable current system was accurate, and the navigation trajectory curve was in good agreement with the theoretical simulation, which effectively verified the correctness of the design of the levitating propulsion system and the coordination of the work of each system. The principal verification of the EDS system has

Figure 5.
Holloman high-speed maglev test device.

Figure 6.
Prototype vehicle of the proof of principle (LTS).

been completed, which lays the foundation for the next full-scale test line construction of the ultra-high speed and low-vacuum pipeline maglev transportation system.

In September 2023, the high-temperature superconducting electric levator navigation test was completed, and a navigation speed of 243 km/h was achieved [13], which is currently the highest high-temperature superconducting electric levator navigation speed recorded in China, as shown in **Figure 7**, the prototype vehicle stop in the brake section when completed the navigation.

In 2022, the General Department of Magneto-Electricity of China Aerospace Science and Industry built a full-size vacuum pipeline test line of about 2 kilometers in Shanxi Province for the joint commissioning test of the whole system.

In January 2023, the Ministry of Magneto-Electricity General Administration of Aerospace Science and Industry completed the first superconducting flight test of the full-size test line (Phase I) of the ultra-high-speed low-vacuum pipeline maglev transportation system at the high-speed Flying Vehicle Test base in Shanxi Province, achieving success [14]. **Figure 8** shows the full-size vehicle and the low-vacuum pipeline. In this test, the full-size superconducting magnet vehicle was used. After completing the cooling and excitation of the superconducting magnet, three navigation tests were completed according to the test procedure. The sailing speed on the 210-meter line was more than 50 km/h. The test is the first full-scale superconducting flight test in China, which verifies key technologies such as high-dynamic superconducting magnets, high-power multiple converters, high-voltage integrated modules, high-vibration suspension framework, high-precision positioning and speed measurement, wireless communication in closed space pipelines, whole-process safety control, and high-precision intelligent rail inspection. The correctness of the overall scheme of the high-speed flying car system and the coordination and matching of the work are preliminarily verified.

In August 2024, the China Academy of Aerospace Science and Industry successfully completed the system integration demonstration and verification test in a low-vacuum environment at the Shanxi high-speed Flying Vehicle test Base [15]. After the establishment of the low-vacuum environment in the 2 km pipeline, the superconducting vehicle set sail, sailed according to the predetermined control curve, stably suspended, and stopped safely. The maximum sailing speed and

Figure 7.
Prototype vehicle of the proof of principle (HTS).

Figure 8.
Full-size vehicle of the proof of principle (HTS).

suspension height were in line with the preset values. All the systems worked normally, and the measured track was in good agreement with the theoretical curve. In this test, for the first time in our country, the full system, full process, and full element of the full-size, high-speed flying vehicle system in low-vacuum environment were realized, and the key technologies of the establishment and maintenance of the long distance and large-size vacuum environment, the superconducting navigation control were verified. The coordination of the various systems and the working performance of the whole system in a low-vacuum environment were verified. The overall technology maturity of the system is further improved, which lays a technical foundation for the subsequent high-speed flying car pilot test verification. **Table 3** summaries the latest development of EDS at the China Academy of Aerospace Science and Industry.

Serial number	age	Test line	vehicle type	Magnet type	Test velocity
1	In 2021.11	400 test line	Prototype vehicle	I-type-LTS	623 km/h
2	In 2023.09	400 test line	Prototype vehicle	I-type-HTS	234 km/h
3	In 2023.01	2 km full-scale line	full-size vehicle	I-type-LTS	50 km/h
4	In 2024.08	2 km full-scale line	full-size vehicle	I-type-LTS	More than 110 km/h

Table 3.
The development of maglev at Aerospace Science and Technology Magneto-electric General Department.

3.4 Maglev in CRRC group, China

In March 2023, the first domestic high-temperature superconducting EDS all-element test system independently developed by CRRC completed its first suspension operation with a sailing speed of 50 km/h [16]. The key core technologies of superconducting maglev transportation system, such as superconducting magnet, linear synchronous traction, inductive power supply, and low-temperature refrigeration, were verified in the suspension operation, which laid a certain foundation for promoting the engineering application of superconducting maglev transportation system. **Figure 9** is the principal prototype of CRRC, which is running in the domestic.

3.5 Maglev in Southwest Jiaotong University, China

In December 2000, the Wang Jiasu team of Southwest Jiaotong University developed the world's first manned high-temperature superconducting pinning suspension car [17].

Figure 9.
Prototype vehicle of the CRRC (HTS).

Figure 10 illustrates the world's first manned high-temperature superconducting pinning suspension car developed by the Wang Jiasu team.

In February 2013, Southwest Jiaotong University developed a 45 m-long vacuum tube high-temperature superconducting napping suspension ring test line with a suspension height of 10 ~ 20 mm and a maximum operating speed of 50 km/h [18]. **Figure 11** shows the vacuum tube guideway (left) and the prototype car (right).

In January 2021, the world's first high-temperature superconducting high-speed maglev engineering prototype vehicle and test line, designed and manufactured by China, were put into operation in Chengdu, with a design speed of 620 kilometers per hour [19], as shown in **Figure 12**, marking a breakthrough in the engineering research of high-temperature superconducting high-speed maglev from scratch, and equipped with the conditions for engineering test and demonstration. The vehicle uses new technologies and processes such as all-carbon fiber lightweight body, low-resistance head type, and large-load high-temperature superconducting maglev technology and is expected to create a new speed record for land traffic in atmospheric environment.

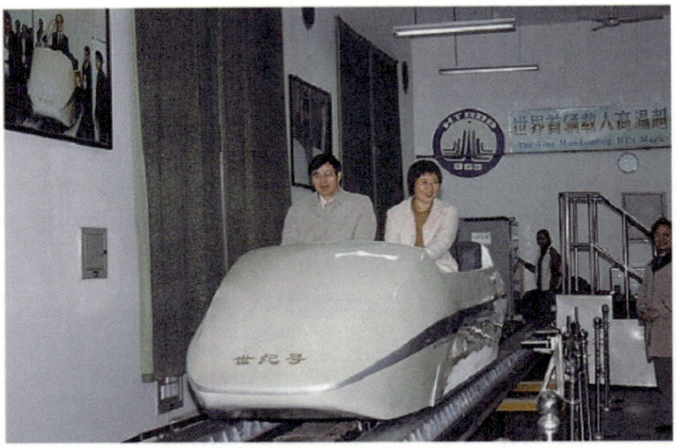

Figure 10.
The world's first manned high-temperature superconducting pinning suspension car.

Figure 11.
Vacuum tube maglev prototype car.

Figure 12.
Engineering prototype car in Southwest Jiaotong University.

4. The key technologies of the superconducting magnet of the EDS

4.1 Miniaturization, lightweight overall design technology

For the ultra-high-speed maglev train, the lighter the overall vehicle is, the more beneficial it is to the high-speed running and levitation of the train. Therefore, the miniaturization and lightweight of the superconducting magnet mounted on the vehicle is an inevitable development trend. At present, in Shanxi full-size test vehicle, the mass of superconducting magnets accounts for 39% of the total test vehicle, and the refrigeration equipment to maintain the operation of superconducting magnets accounts for more than 8%. Magnets and magnetic-related equipment account for nearly half of the mass, which makes the vehicle unable to carry more equipment with other functions and also makes it difficult for the vehicle to achieve the goal of high-speed driving. Therefore, it is very important to develop miniaturized and lightweight magnets.

In order to design miniaturized and lightweight magnets, first of all, we must clarify the design benchmark, components, and functions of magnets. Secondly, find out the parts with larger mass, and reduce the mass without losing its function. In the superconducting magnet, the core components are the superconducting coil and inner dewar, and they are also the largest parts of the mass. Therefore, the miniaturization and lightweight of the superconducting magnet mainly focus on the superconducting coil and inner dewar. The size and quality of superconducting coils are usually designed based on the ampere-turns, centerline size, and the parameters of superconducting wires. Because the ampere-turns and centerline size of the coil are related to the overall performance of the superconducting linear motor, the lightweight of superconducting coils is mainly optimized from the parameters of the superconducting wire itself, winding process, and assembly process.

In the selection of superconducting wire, NbTi/Cu superconducting wire with copper super ratio between 1.0 and 1.5 and large cross section is usually used, so that the superconducting coil density is lower and the number of turns is less, which can effectively reduce the mass and volume of the coil, but the reduction of copper super ratio is also associated with the rise of the risk of losing superconductor.

Therefore, very fine multi-core superconducting wire should be selected in the selection of wire, and the diameter of the superconducting core wire should be reduced as far as possible to make the heat exported by the copper substrate faster and maintain the stability of the superconducting state. In the winding process, skeleton-free winding technology and vacuum impregnation curing technology are used to remove the stainless-steel skeleton of the traditional superconducting coil, and the mass and volume of the superconducting coil are greatly reduced. In the assembly process, a large number of spacer parts are used to fix between the bareless coil and the inner dewar, so that the size of the inner dewar wraps the coil as tightly as possible, and the strength of the inner dewar is relied on to withstand the cold shrinkage force and electromagnetic force. Through the above methods, the mass and volume of the superconducting coil and the inner dewar can be greatly reduced, which lays the foundation for the miniaturization and lightweight of the magnet.

After determining the size of superconducting coil and inner dewar, other components such as outer dewar, cold screen, main support structure, and so on should try to choose metal materials and fiber composite materials with low density and high strength, such as aluminum alloy, titanium alloy, glass fiber composite materials, and so on. Some commonly used materials and their characteristic parameters are shown in **Table 4**.

4.2 Mechanical and electromagnetic interference source blocking technology

In order to realize the miniaturization and light weight of the magnet, the superconducting coil is wound with low copper super ratio wire, which will lead to poor stability of the superconducting coil and prone to quench. Therefore, it is necessary to carry out key technical research on the quench mechanism of the magnet to reduce the risk of quench.

Usually, the reasons for the loss of superconducting coil include temperature rise, increase of magnetic field, increase of current, and so on, because the magnetic field and current of the magnet are in a stable state during operation; the variable magnetic field generated by the ground module coil is usually in the order of tens of mT, which

Serial number	materials	Strength of yield MPa			Thermal conductivity W/(m·K)			density kg/m3
		300 K	77 K	4.2 K	300 K	77 K	4.2 K	
1	Titanium alloyTC4	940	1461	1634	7.5	3.5	0.37	4430
2	Stainless steel 304	200	450	580	15	7.9	0.24	7860
3	Stainless steel 316	240	600	700	14.7	7.9	0.28	7970
4	Aluminum alloy 5083	140	161	178	119.3	55	3.3	2660
5	Aluminum alloy 6061	221	321	368	155.5	85.7	5.4	2700
6	G10	360	600	—	0.99	0.5	0.2	2200

Table 4.
Comparison of structural strength, density, and thermal conductivity of different materials.

can be ignored compared with the coil self-field of 5 T. Therefore, the main cause of inducing the coil out of control can be excluded from the change of magnetic field and current, and the temperature of the locked coil itself rises. To this end, through a large number of investigations and tests, we summarized the mechanism of loss of supercharger causing the temperature rise of the coil, as shown in **Figure 13**. In order to reduce the risk of superconductor coil loss, the harmonic magnetic field generated by the ground module and the transmission structure between the vehicle and the superconducting magnet should be optimized from the source.

Firstly, in terms of optimizing the harmonic magnetic field generated by the ground module, the original single-layer three-phase motor with a phase angle of 120° can be changed to the double-layer three-phase motor with a phase angle of 240°. Under the single-layer motor module, the changing magnetic field generated is 3n ± 1 times harmonic magnetic field in space and 3n times harmonic magnetic field in time, where n is an integer that represents the harmonic order. The changing magnetic field generated by the double-layer motor module is a harmonic magnetic field of 6n ± 1 times in space and 6n times in time. The specific motor arrangement is shown in **Figure 14**.

Through numerical calculation, the three-way electromagnetic force received by the superconducting coil under the two motor arrangements and the harmonic number obtained by Fourier decomposition can be obtained, as shown in **Figure 15**. Through the comparison, it can be seen that the electromagnetic force fluctuation amplitude generated by the double-layer motor is small, and the harmonic amplitude is reduced, and the higher harmonics of more than 2 times can be ignored.

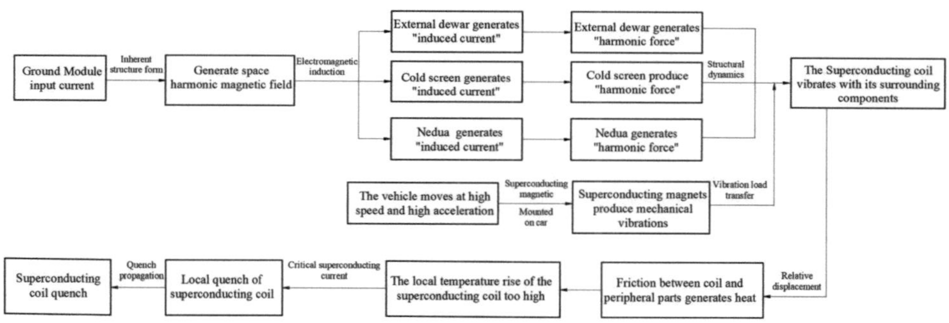

Figure 13.
Electrically suspended superconducting magnets lose supercharge influence chain.

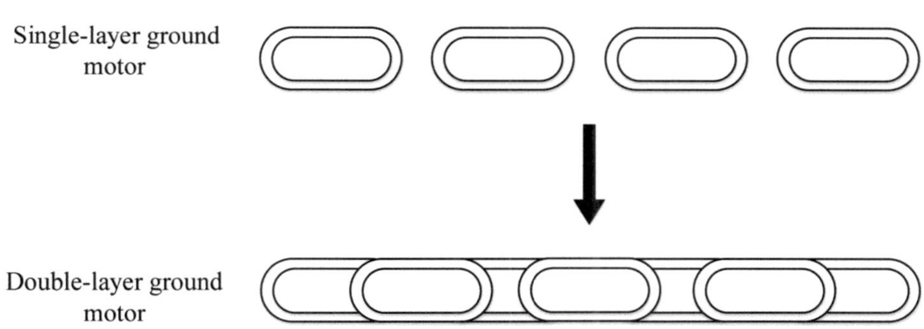

Figure 14.
Ground module structure optimization and promotion.

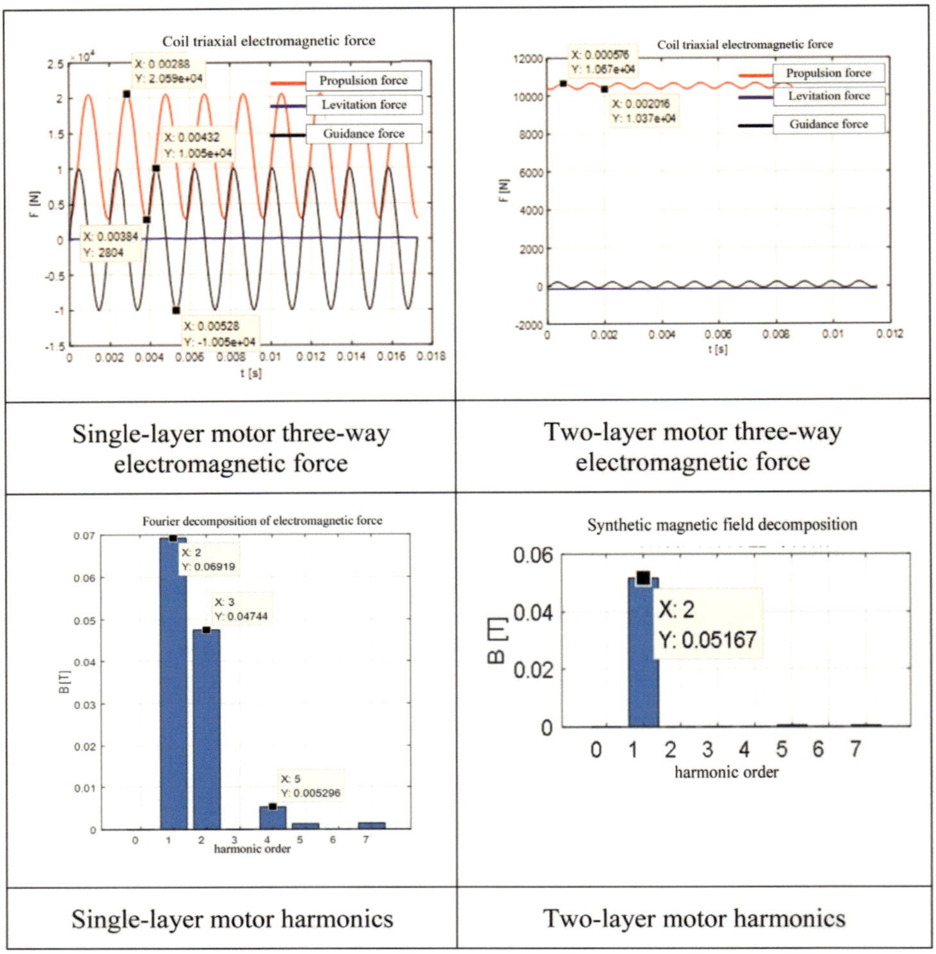

Figure 15.
Comparison of harmonic magnetic fields in single- and double-layer ground motor coils.

After optimizing the coil arrangement of the motor in the ground module, the magnet side can also be further shielded from the changing magnetic field outside the magnet through the skin effect by installing shielding plates with high conductivity materials. Generally, the shielding plate is made of high-electric conductivity aluminum alloy material and installed on the side of the magnet motor. The thickness of the required aluminum shielding plate is calculated by the following penetration formula:

$$\Delta = \sqrt{(2/\omega\mu\gamma)}, \quad (1)$$

where Δ is the penetration depth, ω is the angular frequency ($\omega = 2\pi f$), μ is the magnetic permeability, and γ is the electrical conductivity. The vibration response comparison before and after the use of aluminum shielding plate can be measured by the acceleration sensor installed on the inner dewar. As shown in **Figure 16**, it can be seen that the vibration response is significantly reduced.

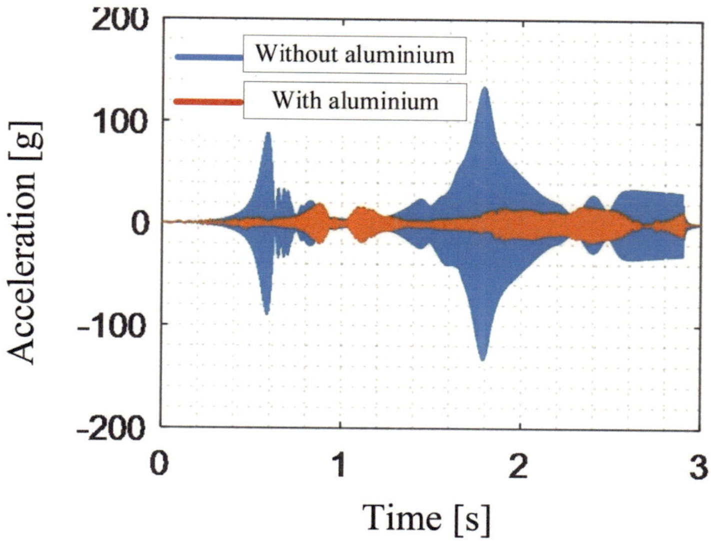

Figure 16.
Comparison of external dewar acceleration response before and after magnetic shielding.

4.3 Technology of improving stability of superconducting magnet in a complex environment

Through the analysis of the mechanism of magnet loss, it can be seen that the main factor causing the coil loss is the temperature rise. Part of the heat resulting in the temperature rise of the coil comes from the electromagnetic heat caused by the harmonic wave of the ground module, and the other part comes from the friction heat generated by the relative friction between the coil and the structure in the vibration environment.

After optimizing the external environment of the magnet, the next step is to optimize the internal structure of the magnet to better cope with the vibration characteristics of the superconducting coil. Taking the magnet of MLX-01 in Japan as an example, when the magnet is sailing along with the train, it will be vibrating by the electromagnetic force from the changing magnetic field of the ground module. Through the modal analysis, the vibration of the coil can be divided into the tumbling mode, torsion mode, bending mode, yawing mode, and so on. Through the mechanical vibration test under the excitation state, the friction heating under various modes is measured in Japan. The support structure of the coil has been continuously optimized, from the initial 4-point support form to the 6-point support form and finally to the 8-point support form, as shown in the following **Figure 17**.

4.4 The large load transfer structure of high-dynamic superconducting coil

In the high-dynamic environment, the strength and heat leakage of the support of the magnet have always been a contradiction. It is necessary to keep the heat leakage from room temperature to liquid helium temperature within the cooling capacity of the liquid helium refrigerator on the premise of ensuring stiffness and strength. Therefore, the heat transfer design and material selection of the support structure should take into account the characteristics of high stiffness and low heat leakage.

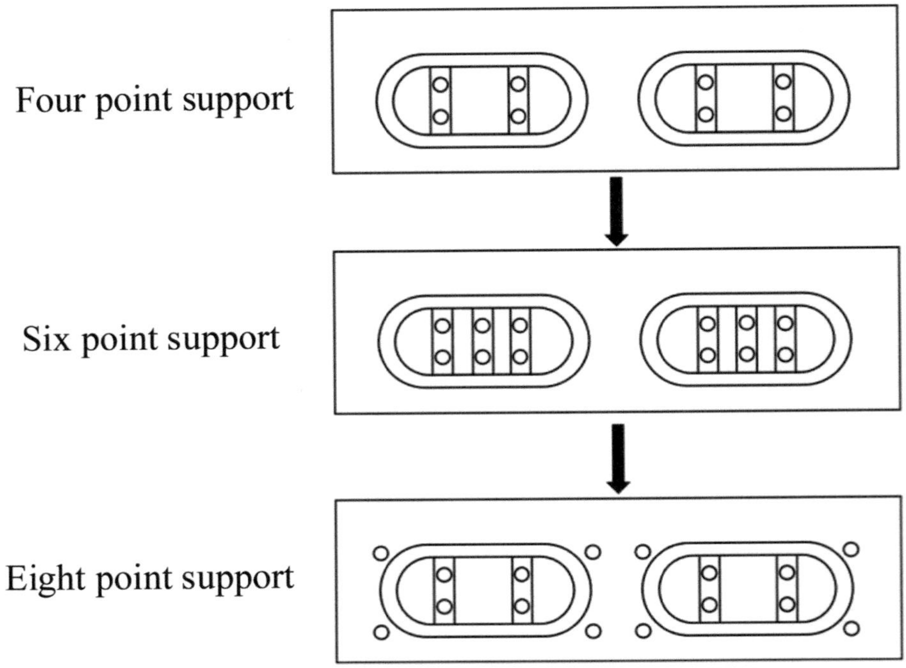

Figure 17.
Superconducting magnet support scheme changes.

Combined with the above two requirements, low thermal conductivity materials such as glass fiber composites and carbon fiber composites are combined with high-strength materials such as stainless steel and titanium alloy, and the corresponding materials are used according to the different working temperature zones and load environments of each material. At the same time, the layered structure is adopted to extend the heat transfer path between high-temperature zone and low-temperature zone as far as possible. The force and heat transfer paths of the magnet are shown in the following **Figures 18** and **19**.

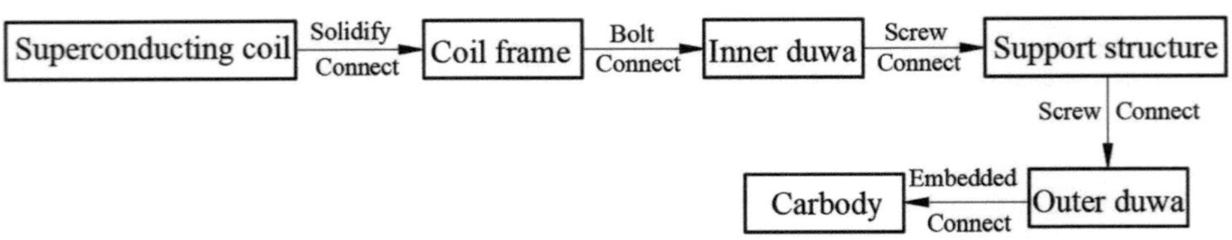

Figure 18.
Superconducting coil electromagnetic force transfer.

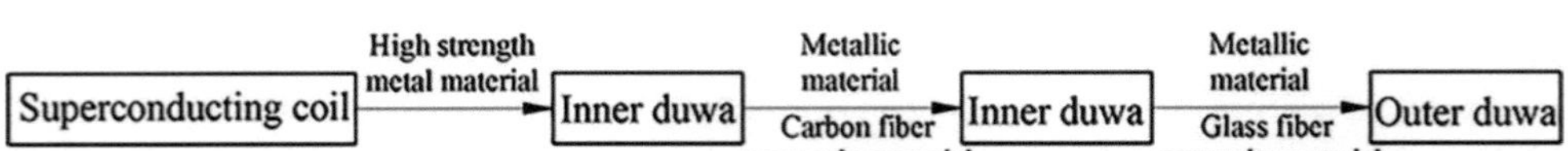

Figure 19.
Magnet heat transfer path.

5. Research prospect of electrically suspended superconducting magnet

The key problems of superconducting magnets for EDS have been solved, and the principal verification, engineering prototype development, and in-line navigation test have been carried out, but there is still a certain gap from the real engineering, and many important problems need to be further studied.

5.1 Chiller/liquid helium closed loop

The heat leakage of the LTS should be further reduced, the cooling capacity of the refrigerator is increased, and the wet closed-loop cooling of the low temperature superconducting magnet liquid helium is realized. And for the HTS, reduce the heat leakage and increase the cooling capacity to realize the closed-loop dry cooling of the refrigerator.

5.2 The temperature margin is higher

The electromagnetic load of superconducting magnets is complex, and it will require long-term stable operation in the future, so it is necessary to develop superconducting wires with high critical performance, optimize the electromagnetic design, improve the temperature margin of superconducting coils, and increase the anti-interference of magnets.

5.3 Lighter and smaller

In order to adapt to the operation scenario of higher speed and higher acceleration and reduce the output power of the suspension propulsion system and the cost of track construction, it is necessary to further reduce the weight and volume of the on-board superconducting magnets and supporting equipment.

5.4 Multi-magnet parallel operation guarantee and low-cost operation

In the future, there will be multiple superconducting magnet systems on trains, so it is necessary to develop a guarantee equipment with multiple magnet simultaneous cooling requirements. Reducing cost is the premise of realizing engineering, carrying out assembly process tooling design, developing liquid helium recovery system, and reducing manufacturing and test costs. **Figure 20** represents the liquid helium

Figure 20.
Liquid helium recycling system.

Figure 21.
Superconducting magnet for EDS in China.

recycling system constructed by Aerospace Science and Technology Magneto-electric General Department China, which can cool a pair of superconducting magnets from 300 k to 4.2 K; it helps reduce the cooling cost of liquid helium.

5.5 Easy maintenance

The fast replacement and convenient maintenance of superconducting magnets are the focus of future research, considering the detachable structure, modularity, and standardization and meeting the requirements of a short maintenance cycle. **Figure 21** illustrates the superconducting magnet used for EDS in China and its undetachable and poor maintenance, which has caused high maintenance cycle and cost in the development.

6. Conclusion

This chapter mainly presents the development status of the superconducting maglev in the world (especially in China), introducing the classification and suspension principle of maglev and describing the research progress and technical challenge of superconducting magnets for EDS. Finally, the research prospect of electrically suspended superconducting magnet in China is proposed.

Acknowledgements

We sincerely express our gratitude to Beijing Natural Science Foundation (Youth Project, No. 3244052). It is precisely due to the strong support and generous funding from Beijing Natural Science Foundation that our research project has been able to proceed smoothly. As a result, valuable progress has been made in exploring the field of superconducting magnetic levitation.

DOI: http://dx.doi.org/10.5772/intechopen.1008167

Author details

Wei Zhou*, Kun Liu, Zhihua Zhang and Si Yuan Liang
Institute of Magnetic Levitation and Electromagnetic Propulsion, China Aerospace Science and Industry Co. Ltd., Beijing, China

*Address all correspondence to: zhouwei910545@126.com

References

[1] Jia-yang X, Zi-gang D. Research progress of high-speed maglev rail transit. Journal of Traffic and Transportation Engineering. 2021;**12**(1):177-198

[2] Shixian L, Lei W, Qiuliang WLW. Review on electrodynamic suspension trains and on-board superconducting magnets. Journal of Southwest Jiaotong University. 2023;**58**(4):734-753

[3] Kolm HH, Thornton RD, Iwasa Y, et al. The magneplane system. Cryogenics. 1975;**15**(7):377-384

[4] Guderjahn CA, Wipf SL, Fink HJ, et al. Magnetic suspension and guidance for high speed rockets by superconducting magnets. Journal of Applied Physics. 1969;**40**(5):2133-2140

[5] Davis LC, Borcherts RH. Superconducting paddle wheels, screws, and other propulsion units for high-speed ground transportation. Journal of Applied Physics. 1973;**44**(7):3294-3299

[6] Coffey H, Solinsky J, Colton J, et al. Dynamic performance of the SRI maglev vehicle. IEEE Transactions on Magnetics. 1974;**10**(3):451-457

[7] Gurol H, Ketchen D, Holland L, et al. Status of the Holloman high speed MagLev test track (HHSMTT). In: 30th AIAA Aerodynamic Measurement Technology and Ground Testing Conference. Atlanta, GA; 2014. p. 2655

[8] Bergeron DA. Holloman High Speed Test Track Maglev Program Update. Nashville: Holloman Air Force Base. (AFB, Rep, AIAA 2010-1707); 2010

[9] Hsu Y, Langhorn A, Ketchen D, et al. Magnetic levitation upgrade to the Holloman high speed test track. IEEE Transactions on Applied Superconductivity. 2009;**19**(3):2074-2077

[10] Hooser C. Proposed hypersonic air breathing test capability at the Holloman high speed test track. In: 22nd AIAA Aerodynamic Measurement Technology and Ground Testing Conference. Louis, Missouri; 2002. pp. 1-12

[11] Bosmajian N, Minto D, Holland L. Status of the magnetic levitation upgrade to Holloman high speed test track. America Institute of Aeronautics & Astronautics; 22 Aug 2012. DOI: 10.2514/6.2000-2289

[12] Maglev Propulsion Test Successful, 623 km/h. Popular Science World. Available from: https://baijiahao.baidu.com/s?id=1715685236018431168&wfr=spider&for=pc

[13] The High Temperature Superconducting Electric Suspension Sailing Test was a Complete Success. Zhongguancun Commercial Space Industry Alliance. Available from: https://mp.weixin.qq.com/s?__biz=Mzg2NDYxMTQyNw= =&mid=2247493068&idx=2&sn=f5d8866d4 dd34191c4ab6ba0e01c631d&chksm=ce6418c 0f91391d67801bfb8fc74d67038 11ee0cf54d1d55b11d22981b 68a645adb2d3035a60&scene=27

[14] The High-Speed Vehicle Completes the First Full-Scale Superconducting Sailing Test. CCTV Finance. Available from: https://v.huanqiu.com/article/4CdJcVxaLaq

[15] Zhu X. Ultra Fast Maglev Train Clears Trial in Shanxi. Available from: https://www.chinadaily.com.cn/

[16] China's First High Temperature Superconducting Electric Suspension Full Element Test System has Completed Its First Suspension Operation. China News Network. Available from: https://china.huanqiu.com/article/4CJJr88PSBI

[17] Wang JS et al. The first man-loading high temperature superconducting Maglev test vehicle in the world. Physica C: Superconductivity. 2002;**378-381**:809-814

[18] Zi-gang D, Wei-hua Z, Jun Z, et al. A high temperature superconducting maglev ring test line developed in Chengdu, China. IEEE Transactions on Applied Superconductivity. 2016;**26**(6):3602408

[19] The World's First High Temperature Superconducting High-Speed Maglev Engineering Sample Car with a Design Speed of 620 km/h Rolled Off the Assembly Line. Xinhua News Agency. Available from: https://www.gov.cn/xinwen/2021-01/13/content_5579614.htm